网络工程师红宝书
思科华为华三实战案例荟萃

周亚军　编著

电子工业出版社
Publishing House of Electronics Industry
北京·BEIJING

内容简介

当今是"互联网+"的时代,许多事物都建立在网络技术之上,很多时候,要求网络工程师熟悉并能应用多个厂商的产品,按需求搭建网络并负责其维护和升级。本书就是为帮助业界同人轻松达到这一目的而编写的。本书是多厂商技术融合的产物,以思科和华为为主,华三为辅,阐述和演示了多个厂商的技术案例实验。对于网络工程师而言,这些案例可以帮助其更好地应付日常工作;对于初学者来说,案例中详尽的命令解释和排错思路可以帮助其更快地入门。本书围绕搭建和维护一个中型企业网络展开实施,可以称之为网络技术常用项目命令大全。作者在书中将在搭建网络和运维网络时最常用的技术分为 39 个案例进行介绍,同时对相关命令做了注解,说明其作用和结果,以帮助读者来理解网络。另外,本书还为读者准备了实验环境(即模拟器)和视频教程,这是实际操作网络设备的重要工具。在读者完成本书学习之后,搭建和维护一个中型企业网也变得简单起来。

本书的目标读者是:没有入门网络技术或者刚刚看到网络技术"大门"的初学者、有一些网络技术基础的人员、稍有项目经验的网络工程师、高等院校相关专业的师生。

未经许可,不得以任何方式复制或抄袭本书之部分或全部内容。
版权所有,侵权必究。

图书在版编目(CIP)数据

网络工程师红宝书:思科华为华三实战案例荟萃 / 周亚军编著. —北京:电子工业出版社,2020.5
ISBN 978-7-121-37662-7

Ⅰ. ①网… Ⅱ. ①周… Ⅲ. ①计算机网络-基本知识 Ⅳ. ①TP393

中国版本图书馆 CIP 数据核字(2019)第 242898 号

责任编辑:宋　梅　　文字编辑:满美希
印　　刷:涿州市般润文化传播有限公司
装　　订:涿州市般润文化传播有限公司
出版发行:电子工业出版社
　　　　　北京市海淀区万寿路 173 信箱　邮编:100036
开　　本:787×1 092　1/16　印张:19.25　字数:493 千字
版　　次:2020 年 5 月第 1 版
印　　次:2025 年 3 月第 14 次印刷
定　　价:78.00 元

凡所购买电子工业出版社图书有缺损问题,请向购买书店调换。若书店售缺,请与本社发行部联系,联系及邮购电话:(010)88254888,88258888。

质量投诉请发邮件至 zlts@phei.com.cn,盗版侵权举报请发邮件至 dbqq@phei.com.cn。
本书咨询联系方式:mariams@phei.com.cn。

关于作者

周亚军（安德）

大部分学员口中的军哥，CCIE、HCIE 讲师中的"段子王"，课上幽默、课下严谨的典范，是所属实验室乾颐堂的劳模讲师，51CTO 金牌讲师，他的良好口碑是通过孜孜不倦的服务和平易近人得来的。

网络技术畅销书作者，著有经典网络技术畅销书《思科 CCIE 路由交换 v5 实验指南》《思科运营商 CCIE 认证实现指南》《华为 HCNA 认证详解与学习指南》，这些书籍对读者学习和掌握网络技术大有帮助。

华为 HCIEv2.0 通过第一人，思科路由交换和运营商互联网专家。在他带领的乾颐堂大路由组中，通过 IE 认证（思科 CCIE 认证和华为 HCIE 认证）的学员超过 700 人，其他的 NA 级别、NP 级别学员不计其数。他拥有自己的思科华为实验室，以专业的教育方式培养了大批专家级学员，以 2018 年为例，乾颐堂实验室有 294 名学员通过 IE 认证，其中大路由组通过 151 名。

周亚军老师主持的众多大型企业培训名录：神华集团有限责任公司（世界 500 强企业）、国家电网有限公司（世界 500 强企业）、中国联通有限公司上海分公司、索尼（中国）有限公司无锡分公司、思科（世界 500 强企业）上海区下一代 CCNP 网络技术培训、思科"思蜀援川"项目。周亚军老师曾担任某驻京部队网络技能比武导师（其指导的学员获得金牌），同时，他还在上海电子信息职业技术学院任教。

关于周亚军老师的更多信息，读者可以从百度百科了解，链接：https://baike.baidu.com/item/周亚军，或者在百度百科中搜索周亚军老师的名字，选择"互联网讲师"。

致 谢

  首先,要感谢电子工业出版社的宋梅老师不辞辛苦的工作,才使本书得以出版,感激之情溢于言表。我们的共同目的是普及网络的系统知识。

  其次,要感谢我的夫人,能够容忍我在半夜写书的时候打扰她和孩子,能够容忍叫我三遍后才吃午饭,能够容忍我因为思考技术问题走神而摔倒了还是幼儿的孩子,能够容忍我每周休息1天而没时间陪伴家人。

前　言

本书内容和结构

整体而言，本书是多厂商技术融合的产物，以思科和华为为主，华三为辅，阐述和演示多个厂商的技术案例实验。对于网络工程师而言，这些案例可以帮助其更好地应付日常工作；对于初学者来说，案例中详尽的命令解释和排错思路可以帮助其更快地入门。

在当今的"互联网+"时代，社会上的政企单位更多时候要求网络工程师具备掌握"多厂商"网络设备的能力，而"多厂商"在国内具体指思科、华为和华三这三家著名企业。当然了，之所以是这三家企业是有一定道理的。思科是网络技术界的鼻祖，也是世界500强的老牌企业。华为是我国重要的民族企业，过去主要涉足运营商网络，而近五六年开始进入企业网络领域，品牌的力量不可小觑。而华三虽然成长过程跌宕起伏，但是由于长期服务于广大企业网，也积累了相当好的口碑。本书迎合时代的需求，通过众多的案例来帮助读者理解和实现网络，而这些案例就来源于这三家企业的平台。

本书的实操性非常强，同时保留了理论部分，这一部分被称为"简书"，在每部分知识中都包含对应的理论简书。当然，想要了解更多理论内容，读者可以参阅视频课程和互联网资料。笔者已经为读者准备了课前资料，这里包含了几个厂商的模拟器和必需的软件，当然也包含了部分笔者的课程视频资料。（链接：https://pan.baidu.com/s/1yEf7UUigji1 BpdUkmScakw，提取码：9x31。）如果链接失效，请读者联系乾颐堂实验室（www.qytang.com），或者加入QQ群（515840631）参与讨论。

本书分为网络实施基础、交换网络、路由网络以及网络扩展与公网接入4篇。

第一篇为读者阐述如何使用当下流行的学习模拟器、多种网络工程师必备软件的部署、IP地址配置、本地和远程管理网络设备、对网络设备进行文件管理等基础内容的实施。本篇内容是网络工程师经常需要的内容，也是必须掌握的内容。

第二篇介绍网络工程师必备的交换网络知识，为读者演示了MAC地址表、VLAN接入、接入端口、干道模式和混杂端口、生成树协议、以太网链路聚合、交换机堆叠、VRRP实施、VLAN间路由通信以及端口安全等技术。掌握该部分知识后，就算基本入门网络技术了，当然掌握是指灵活掌握。交换网络知识的特点是相对零碎，需要有较强的逻辑性，与路由网络知识相比更实用。本篇内容更多地被应用于接入网络和汇聚网络，当然也包含汇聚网络到核心网络的接入。

第三篇介绍的路由网络技术扩展了网络的边界，可以把用户的业务转发到更远的网络。本篇内容包括路由技术原理（请注意第二篇中的VRRP、VLAN间路由在一定程度上属于直连路由的内容）、企业网中使用最多的OSPF协议、GRE隧道封装以及城域网中经常使用的非常关键的BGP（BGP也是NP和IE级别认证中最重要的一个协议）。在本篇中，并未介绍以往书籍中经常讲到的 RIP 和 EIGRP。就 RIP 而言，这个协议确实过于"古老"，在现实网络中几乎绝迹，当然 NP 和 IE 级别考试对此也几乎不会涉及，所以本书没有对其进行介绍（如果想了解相关内容，可以观看视频资料，视频资料中讲解了这部分内容）。而对于 EIGRP，不可否认这是一个非常好的 IGP，而且在很多金融企业的现实网络中都对其有所涉及，但由于它是思科私有的路由协议，因此也没有放入本书中讲解，但在笔者的 NP 课程体系中对其进行了详细的讲解，读者可关注笔者的其他书籍和课程。

第四篇介绍网络扩展与公网接入相关内容，该篇内容对于网络安全管理、网络质量管理、互联网接入是必备知识。具体内容包括访问控制列表、网络地址转换（NAT）、网络服务质量（QoS）限速和 PPPoE 技术。

本书围绕搭建和维护一个中型企业网络展开实施，在读者完成本书的学习之后，构建和维护一个中型企业网络也会变得简单起来。

本书适合读者和笔者心语

本书面向的读者更倾向于没有入门网络技术或刚刚看到网络技术"大门"的初学者，同时本书对于已经有一些基础、稍有项目经验的网络工程师亦有帮助。在这里我们可能需要稍微讨论一下学习方法和学习效率的问题，虽然笔者一直致力于编写各种图书，但一定要指出的是，网络技术是一门应用技术学科，而不是学术性学科，这意味着动手实操一定要优先于一味地"看死书""死看书"，而且很多图书动辄近千页，很多时候不等读者入门，耐心可能就已经被消耗殆尽了。怎么才能更快地入门和学习呢？笔者需要一吐 10 年教学经验，读者可以酌情选择以下学习方法。

① 通过直播课程学习。实时提问和解答的互动方式可以提高学习效率。笔者所在实验室已经坚持线上、线下教育 5 年，并培养出数十万名网络工程师，欢迎大家一同来学习。读者可以在互联网搜索"乾颐堂安德"或"乾颐堂 QCNA 课程"找到直播课程。

② 坚持做作业和实验。学以致用才能加速入门和理解技术，笔者在每次教学中都会留动手实操的作业，每次课前也会进行课前测试以便学员加深理解。不断地实验是学习的最好途径。

③ 不耻下问。课下遇到问题要在 QQ 群内积极提问，老师或同学的帮助会让人"恍然大悟"。当然，之前录制的视频有助于很好地预习和复习。

④ 笔者独家撰写的课件要比官方文档更加简洁而且更有针对性。抓住重点学习可以提高学习效率。

⑤ 课下学习参考书。请参考笔者编写的网络类书籍，海量实验和详细命令解析、清晰的实施步骤和验证更便于培养实施和排障思路。本书也是这其中之一。

⑥ 英语并不是考试的拦路虎，网络是应用学科，考生要做的只是熟悉理论并实现理论的命令而已。当然，英语好的话对日后找工作以及跳槽的帮助非常大。

⑦ 关于学习周期的问题，每个人的目标不同，努力程度以及理解能力也大相径庭，坚持和努力才是最好的学习方式。请问读者坚持完成以上建议了吗？当然了，站在一个普遍的角度看，IE 级别的学习周期在 4 个月到 7 个月之间。

笔者是一名从事多年网络技术教学和认证考试培训的讲师，更多的时候是一名服务人员，服务于广大的学员（学员中从事多年专业网络项目的工程师居多，还有一部分学员是转行人员，从其他行业比如机床操作工、理发师、设计师等转行来学习网络技术，当然也有许多精力充沛的大学生）。笔者对学员们苦口婆心地叮嘱：请一遍遍地通过实验学习网络技术，在错误中成长为成熟的网络工程师（当然，或许读者更需要一张证明能力的 CCIE 或者 HCIE 认证证书）。如果读者充分利用好本书，至少能应付初、中级的网络项目工程。

本书是乾颐堂独家融合思科 CCNA 和华为 HCIA 课程（以及部分华三课程）内容（命名为 QCNA，即全方向 NA）的辅导书籍，更多内容请关注笔者的直播课程。

目　录

第一篇　网络实施基础

案例 1　模拟器的部署和连接管理 ... 3
　1.1　华为模拟器 eNSP 部署 ... 3
　1.2　思科模拟器 EVE 部署 ... 8
　1.3　部署 SecureCRT 管理网络设备 ... 13
　　1.3.1　部署终端管理软件管理华为模拟设备 ... 13
　　1.3.2　部署终端管理软件管理思科模拟设备 ... 15

案例 2　网络设备初始化及 Console 端口密码认证 ... 17
　2.1　思科设备初始化及 Console 端口密码认证 ... 19
　2.2　华为设备初始化及 Console 端口密码认证 ... 20
　2.3　华三设备初始化及 Console 端口密码认证 ... 22

案例 3　配置明文远程管理协议 ... 24
　3.1　在思科设备上配置远程管理协议 ... 24
　　3.1.1　通过 Telnet 协议远程管理思科网络设备 ... 24
　　3.1.2　配置 Telnet 协议拓扑说明 ... 25
　　3.1.3　远程管理配置要点 ... 25
　　3.1.4　协议配置步骤 ... 25
　3.2　在华为设备上配置远程管理协议 ... 26
　　3.2.1　通过 Telnet 协议远程管理华为网络设备 ... 26
　　3.2.2　配置远程管理协议拓扑说明 ... 26
　　3.2.3　远程管理配置要点 ... 27
　　3.2.4　远程管理配置详解 ... 27
　3.3　在华三设备上配置远程管理协议 ... 29
　　3.3.1　通过 Telnet 协议远程管理华三网络设备 ... 29
　　3.3.2　华三设备远程管理协议配置拓扑 ... 29
　　3.3.3　远程管理配置要点 ... 29
　　3.3.4　远程管理配置详解 ... 29

案例 4　配置 SSH 协议 ... 32
　4.1　在思科设备上配置 SSH 协议 ... 32
　　4.1.1　客户安全管理需求 ... 32

  4.1.2 使用 SSH 协议组网拓扑 ·················· 32
  4.1.3 配置 SSH 协议要点 ···················· 33
  4.1.4 配置 SSH 协议步骤详解 ·················· 33
 4.2 在华为设备上配置 SSH 协议 ·················· 34
  4.2.1 使用 SSH 协议的组网需求 ················· 34
  4.2.2 使用 SSH 协议的组网拓扑 ················· 35
  4.2.3 配置 SSH 协议要点 ···················· 35
  4.2.4 配置 SSH 协议步骤详解 ·················· 35
 4.3 在华三设备上配置 SSH 协议 ·················· 37
  4.3.1 安全管理网元需求 ····················· 37
  4.3.2 使用 SSH 协议组网拓扑 ·················· 37
  4.3.3 配置 SSH 协议要点 ···················· 38
  4.3.4 配置 SSH 协议和测试步骤 ················· 38

案例 5 配置 TFTP ······························ 39
 5.1 在思科设备上配置 TFTP ···················· 39
  5.1.1 使用 TFTP 的组网需求 ·················· 39
  5.1.2 使用 TFTP 的组网拓扑 ·················· 39
  5.1.3 配置 TFTP 要点 ····················· 39
  5.1.4 配置 TFTP 步骤详解 ··················· 40
 5.2 在华为及华三设备上实现 TFTP ················ 41
  5.2.1 使用 TFTP 的组网需求 ·················· 41
  5.2.2 使用 TFTP 的组网环境 ·················· 41
  5.2.3 配置 TFTP 要点 ····················· 41
  5.2.4 配置 TFTP 步骤详解 ··················· 41

案例 6 配置 FTP ······························· 44
 6.1 配置 FTP 案例拓扑 ······················· 44
 6.2 使用 FTP 的组网需求 ····················· 45
 6.3 配置 FTP 步骤详解 ······················ 45
  6.3.1 配置路由器基本网络和 FTP 服务 ············· 45
  6.3.2 配置 FTP 客户端 ····················· 46

案例 7 网络设备文件系统管理 ······················ 48
 7.1 文件系统管理案例拓扑 ···················· 48
 7.2 文件管理需求 ························· 48
 7.3 文件管理系统配置步骤详解 ·················· 49
  7.3.1 熟悉 VRP 系统的基本查看命令 ·············· 49
  7.3.2 配置 VRP 系统的目录 ·················· 50
  7.3.3 配置 VRP 系统的文件 ·················· 50

7.3.4	管理 VRP 系统的配置文件	52
7.3.5	指定 VRP 系统启动文件和恢复出厂配置	55
7.3.6	熟悉思科网元的文件系统操作	57

第二篇 交换网络

案例 8	交换机的 MAC 地址表	61
8.1	查验 MAC 地址表网络拓扑	62
8.2	学习 MAC 地址表的现实需求	62
8.3	验证 MAC 地址表的要点	62
8.4	验证 MAC 地址表的实施步骤	63
8.4.1	在华三设备上学习 MAC 地址表	63
8.4.2	在思科设备上学习 MAC 地址表	65
8.4.3	在华为设备上学习 MAC 地址表	66
案例 9	VLAN 技术接入案例	70
9.1	实施端口接入 VLAN 的组网拓扑	72
9.2	实施端口接入 VLAN 的组网需求	72
9.3	VLAN 接入技术的配置要点	72
9.4	VLAN 接入技术实施步骤详解	72
9.4.1	在华为设备上实施 VLAN 接入	72
9.4.2	在思科设备上实施 VLAN 接入	75
9.4.3	在华三设备上实施 VLAN 接入	77
案例 10	交换机互连链路 VLAN 接入模式	80
10.1	交换机互连链路配置 VLAN 接入模式描述	81
10.2	组网拓扑介绍	81
10.3	配置要点	81
10.4	交换机互连链路 VLAN 接入模式配置详解	81
10.4.1	华为交换机互连链路 VLAN 接入模式配置	81
10.4.2	思科交换机互连链路 VLAN 接入模式配置	84
10.4.3	华三交换机互连链路 VLAN 接入模式配置	86
案例 11	交换机的 Trunk（干道）模式	88
11.1	Trunk 模式应用场景	89
11.2	Trunk 模式配置案例拓扑说明	89
11.3	Trunk 模式配置要点	89
11.4	Trunk 模式配置详解	90
11.4.1	在华为设备上配置 Trunk 模式	90

| | 11.4.2 | 在思科设备上配置 Trunk 模式 | 92 |
| | 11.4.3 | 在华三设备上配置 Trunk 模式 | 93 |

案例 12　Trunk 上本征 VLAN 或 PVID 的最佳实践 94

12.1	调整 Trunk 上的本征 VLAN 或 PVID	95
12.2	实施拓扑说明	95
12.3	调整 Trunk 上本征 VLAN 或 PVID 的配置要点	95
12.4	调整 Trunk 上本征 VLAN 或 PVID 的配置详解	95
	12.4.1　在华为设备的 Trunk 链路上调整 PVID VLAN	95
	12.4.2　在思科设备的 Trunk 链路上调整本征 VLAN	97

案例 13　华为交换机上的 Hybrid（混杂）端口 100

13.1	交换机混杂模式应用场景	101
13.2	交换机混杂模式配置拓扑说明	101
13.3	混杂模式配置要点	101
13.4	混杂模式配置步骤详解	101
	13.4.1　将交换机互连端口配置为混杂模式	101
	13.4.2　将连接终端的端口配置为混杂模式	104

案例 14　华为华三的标准生成树协议 109

14.1	在华为华三网络设备上配置标准生成树协议	118
14.2	配置生成树协议要点	119
14.3	配置标准生成树协议详解	119
	14.3.1　修改生成树模式及生成树的根和备份根	120
	14.3.2　在接入层交换机上配置边缘端口	123
	14.3.3　在接入层交换机上配置 BPDU 保护功能	125
	14.3.4　华三设备生成树配置命令参考	127

案例 15　思科的标准生成树协议 129

15.1	在思科设备上配置标准生成树协议	129
15.2	案例拓扑说明	130
15.3	配置生成树协议要点	130
15.4	配置标准生成树协议步骤详解	130
	15.4.1　调整生成树模式及根设备	130
	15.4.2　配置边缘端口	131
	15.4.3　配置 BPDU 保护功能	133

案例 16　华为设备以太网链路聚合 135

16.1	华为设备以太网链路聚合案例说明	137
16.2	华为设备以太网链路聚合实施拓扑	137
16.3	华为设备配置以太网链路聚合要点	138
16.4	华为设备配置以太网链路聚合步骤详解	138

	16.4.1 华为设备以太网链路聚合的前置条件和基本配置	138
	16.4.2 以太网聚合端口在 STP 中的状态和 Trunk 模式配置	139
	16.4.3 华三设备链路聚合命令参考	140

案例 17 思科设备以太网链路聚合 ························ 142
   17.1 思科设备以太网链路聚合实施拓扑 ························ 142
   17.2 思科设备以太网链路聚合配置要点 ························ 142
   17.3 思科设备以太网链路聚合配置步骤详解 ···················· 142

案例 18 交换机堆叠技术 ·· 145
   18.1 交换机堆叠案例实施说明 ·· 147
   18.2 实施交换机堆叠拓扑 ·· 148
   18.3 交换机堆叠配置要点 ·· 148
   18.4 交换机堆叠配置步骤详解 ·· 148
      18.4.1 配置 IRF 的 Master 设备优先级并加入物理端口 ······· 148
      18.4.2 配置 IRF 的 Slave 设备板卡序号 ······························· 149
      18.4.3 激活 IRF 的配置方法 ·· 149

案例 19 华为华三设备的单臂路由技术 ················· 151
   19.1 单臂路由实施案例及组网拓扑 ·································· 152
   19.2 单臂路由配置要点 ·· 152
   19.3 配置单臂路由步骤详解 ·· 153
      19.3.1 在路由器上配置子接口，标识特定 VLAN 的流量 ······· 153
      19.3.2 在交换机上配置 Trunk 和 Access 链路 ························ 154
      19.3.3 配置 PC 的地址并完成数据测试 ································· 155
      19.3.4 华三设备实现单臂路由命令参考 ································· 157

案例 20 在思科设备上配置单臂路由 ······················· 158
   20.1 在思科设备上配置单臂路由案例 ······························ 158
   20.2 案例拓扑说明 ·· 158
   20.3 在思科设备上配置单臂路由步骤详解 ······················ 159
      20.3.1 在路由器上配置子接口，标识特定 VLAN 的流量 ······· 159
      20.3.2 在思科交换机上配置 Trunk 和 Access 链路 ················ 159
      20.3.3 完成终端配置和通信结果测试 ····································· 160

案例 21 华为华三网元的 VLANIF ······························ 162
   21.1 VLANIF 实施案例及拓扑 ·· 163
   21.2 VLANIF 配置要点 ·· 163
   21.3 配置 VLANIF 步骤详解 ··· 164
      21.3.1 配置华为设备上的 VLANIF 接口 ······························ 164
      21.3.2 华为设备数据测试和华三命令参考 ···························· 165

案例 22　思科网元的 SVI 技术 ·· 167
22.1　使用 SVI 技术组网案例及实施拓扑 ·· 167
22.2　SVI 技术配置要点 ·· 167
22.3　配置 SVI 步骤详解 ··· 167
22.3.1　配置 SVI 的前置条件 ·· 167
22.3.2　在交换机上配置 SVI ··· 168
22.3.3　配置路由器模拟的终端并测试数据通信 ····························· 169

案例 23　二层交换接口转换为三层路由接口的方案 ·································· 171
23.1　交换机的三层路由接口实施案例 ·· 171
23.2　交换机的三层路由接口实施拓扑 ·· 172
23.3　交换机三层接口配置要点 ·· 172
23.4　交换机三层接口配置步骤 ·· 172
23.4.1　在思科设备上配置交换机三层接口 ································· 172
23.4.2　在华为设备上实施交换机三层接口 ································· 173

案例 24　在华为设备上配置 VRRP ·· 174
24.1　VRRP 实施案例 ·· 176
24.2　VRRP 案例拓扑 ·· 176
24.3　VRRP 配置要点 ·· 177
24.4　配置 VRRP 步骤详解 ··· 177
24.4.1　在交换机上配置直连的 VLANIF 地址 ······························ 177
24.4.2　在 VLAN8 中配置 VRRP ··· 178
24.4.3　在其他 VLAN 中配置 VRRP ······································ 179

案例 25　在思科设备上配置 VRRP ·· 181
25.1　VRRP 实施案例及拓扑 ·· 181
25.2　VRRP 配置要点 ·· 181
25.3　VRRP 配置步骤 ·· 182

案例 26　在华为设备上实施端口安全技术 ·· 184
26.1　实施端口安全技术的拓扑 ·· 184
26.2　端口安全的配置要点 ·· 184
26.3　配置端口安全步骤详解 ·· 185
26.3.1　在交换机连接终端的端口上实施端口安全技术 ······················· 185
26.3.2　验证违规行为 ·· 186
26.3.3　华三设备实施端口安全参考命令 ··································· 186

案例 27　在思科设备上实施端口安全技术 ·· 188
27.1　实施端口安全技术的拓扑 ·· 188
27.2　端口安全的配置要点 ·· 188
27.3　配置端口安全步骤详解 ·· 189

27.3.1 在交换机上实施端口安全技术 189
27.3.2 验证端口安全的违规行为和自动恢复功能 189

第三篇 路由网络

案例 28 在华为设备上配置静态路由 193
28.1 静态路由应用案例及拓扑 195
28.2 配置静态路由要点 196
28.3 配置静态路由步骤详解 196
    28.3.1 在网关设备上配置静态明细路由 196
    28.3.2 在网关设备上配置浮动静态路由 197
    28.3.3 在网关设备上配置浮动静态默认路由 199

案例 29 在思科设备上配置静态路由 203
29.1 静态路由配置案例及拓扑 203
29.2 静态路由配置要点 203
29.3 静态路由配置步骤详解 204
    29.3.1 在网关设备上配置静态明细路由 204
    29.3.2 在网关设备上配置浮动静态路由 205
    29.3.3 在网关设备上配置浮动静态默认路由 207

案例 30 在华为设备上配置 OSPF 209
30.1 在企业内部网络上配置 OSPF 211
30.2 配置 OSPF 的要点 211
30.3 配置 OSPF 的步骤详解 212
    30.3.1 在重要的网络设备之间配置 OSPF 212
    30.3.2 修改 OSPF 的网络类型 218
    30.3.3 在企业网关设备上向其他交换机下发默认路由 220

案例 31 在思科设备上配置 OSPF 222
31.1 在企业内部网络上配置 OSPF 222
31.2 配置 OSPF 的要点 222
31.3 配置 OSPF 的步骤详解 223
    31.3.1 配置基本的 OSPF 邻居并在业务网络中运行 OSPF 223
    31.3.2 在企业网关设备上向局域网下发 OSPF 默认路由 227

案例 32 在华为设备上配置 GRE 隧道和 EBGP 230
32.1 GRE 协议和 EBGP 应用案例说明 231
32.2 GRE 协议和 EBGP 应用案例拓扑 231
32.3 GRE 隧道和基础 BGP 配置要点 232
32.4 GRE 隧道和基础 BGP 配置步骤详解 232

32.4.1 在华为设备上配置 GRE 隧道 232
   32.4.2 通过配置 EBGP 邻居实现站点间通信 235

案例 33 在思科设备上配置 GRE 隧道和 EBGP 240
 33.1 GRE 协议和 EBGP 应用案例说明 240
 33.2 GRE 协议和 EBGP 应用案例拓扑 240
 33.3 GRE 隧道和基础 BGP 配置要点 240
 33.4 GRE 隧道和基础 BGP 配置步骤详解 241
   33.4.1 配置穿越互联网的 GRE 隧道 241
   33.4.2 使用隧道的私网地址建立 EBGP 邻居并通信 243

第四篇 网络扩展与公网接入

案例 34 在华为设备上配置访问控制列表 249
 34.1 配置访问控制列表案例及拓扑 250
 34.2 配置 ACL 要点 250
 34.3 配置 ACL 步骤详解 250
   34.3.1 在接口上配置基本 ACL 用于过滤流量 250
   34.3.2 验证 ACL 管理设备自身发起的流量 253
   34.3.3 将高级 ACL 应用于接口管理流量 254
   34.3.4 将 ACL 应用于用户接口实现远程管理 255

案例 35 在思科设备上配置访问控制列表 257
 35.1 配置访问控制列表案例及拓扑 257
 35.2 配置 ACL 要点 257
 35.3 配置 ACL 步骤详解 258
   35.3.1 在接口上配置标准 ACL 用于过滤流量 258
   35.3.2 验证 ACL 管理设备自身发起的流量 259
   35.3.3 将扩展 ACL 应用于接口管理流量 259
   35.3.4 将 ACL 应用于 VTY 接口实现远程管理 260

案例 36 在华为设备上配置 NAT 功能 262
 36.1 配置 NAT 功能案例及拓扑 262
 36.2 配置 NAT 功能要点 262
 36.3 配置 NAT 功能步骤详解 263
   36.3.1 配置 Easy IP 功能实现内网访问互联网 263
   36.3.2 配置 NAT 服务器功能向互联网提供服务 265

案例 37 在思科设备上配置 NAT 功能 268
 37.1 配置 NAT 功能案例 268
 37.2 配置 NAT 功能案例拓扑 268

	37.3	配置 NAT 功能要点	268
	37.4	配置 NAT 功能步骤详解	269
		37.4.1 配置 PAT 功能实现对应内网访问互联网业务	269
		37.4.2 配置静态 NAT 功能以对外提供服务	270

案例 38　在华为设备上实施网络 QoS 限速 272
- 38.1 实现企业网络限速管控案例及拓扑 274
- 38.2 配置 QoS 要点 274
- 38.3 配置 QoS 步骤详解 274
  - 38.3.1 在华为设备上实施监管功能以实现限速 274
  - 38.3.2 在华为设备上实施流量整形功能以实现限速 276

案例 39　在思科设备上实施网络 QoS 限速 279
- 39.1 实现企业网络限速管控案例及拓扑 279
- 39.2 配置 QoS 要点 279
- 39.3 配置 QoS 步骤详解 280
  - 39.3.1 在思科设备上实施监管功能以实现限速 280
  - 39.3.2 在思科设备上实施流量整形功能以实现限速 281

术语表 283

第一篇　网络实施基础

本篇将为读者展现网络学习和应用中最基本的部分，这些内容包括：模拟器的部署和连接管理；Console（控制台）基础和使用方法；IP 地址配置，以及明文和密文远程管理网络设备；FTP 在管理网络设备时的应用；TFTP 在管理网络设备时的应用；网络设备的文件系统管理。这些内容都是网络工程师在日常工作中会用到的重要的基础应用。

本篇内容包含案例 1~7。

第一篇 网络文化基础

案例 1　模拟器的部署和连接管理

学习利器——模拟器简书

对于网络技术初学者来说，有两个要素会影响其是否可以真正入门。这两个要素是：① 能否找到一个实用的模拟器；② 能否找到一个领路人指导其学习（虽然有很多很好的书籍可供自学，但自学是一件痛苦而漫长，同时又效率极低的方式）。笔者推荐两款当下最为流行的模拟器，思科模拟器 EVE 和华为模拟器 eNSP（这两款模拟器是思科、华为 IE 级别认证考试中使用的环境）。笔者已经将这两款模拟器分享到百度网盘，下载链接：https://pan.baidu.com/s/1KxUBd0B3QYLqSLm5E_Oqew，提取码：3pc5。（若下载链接失效，请到乾颐堂官网（www.qytang.com）或到笔者博客 http://blog.51cto.com/enderjoe 留言获得。）

1.1　华为模拟器 eNSP 部署

下载完模拟器安装包后，选中所有的插件，然后遵循单击"下一步"的安装原则即可。关于该款模拟器推荐使用 eNSP：V100R002C00B500 版本（即我们通常所说的 500 版本 eNSP，并不推荐使用较新的 510 版本，该版本容易出现配置丢失、某些设备不能启动、NAT 无效等问题）。

安装 eNSP 模拟器的步骤如下。

第 1 步，如图 1-1 所示，双击安装程序，安装 eNSP。

第 2 步，如图 1-2 所示，选择接受用户协议，然后单击"下一步(N)"按钮。

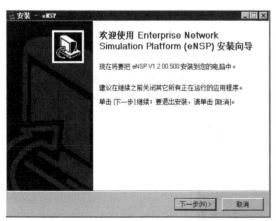
图 1-1　双击安装程序，安装 eNSP

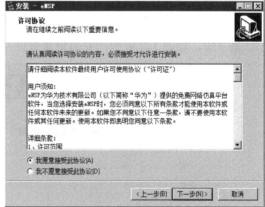
图 1-2　选择接受用户协议

第 3 步，如图 1-3 所示，选择 eNSP 安装目录，请使用英文目录，然后单击"下一步(N)"按钮。

第 4 步，如图 1-4 所示，安装 eNSP 的开始菜单选项，使用默认选项，然后单击"下一步(N)"按钮。

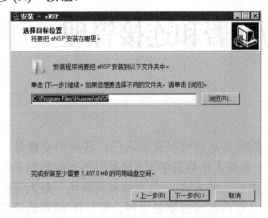 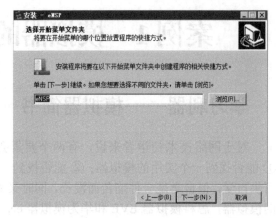

图 1-3　选择 eNSP 安装目录　　　　　　图 1-4　安装 eNSP 的开始菜单选项

第 5 步，如图 1-5 所示，安装附加任务，使用默认选项，然后单击"下一步(N)"按钮。
第 6 步，如图 1-6 所示，选择安装所有软件，然后单击"下一步(N)"按钮。

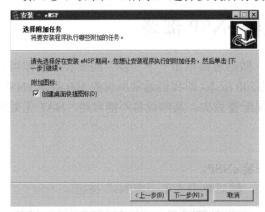 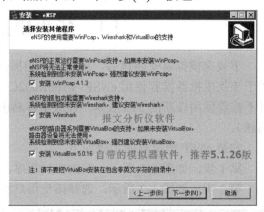

图 1-5　安装附加任务　　　　　　　　　图 1-6　选择安装所有软件

第 7 步，如图 1-7 所示，安装附属软件 WinPcap，然后单击"Next"按钮。

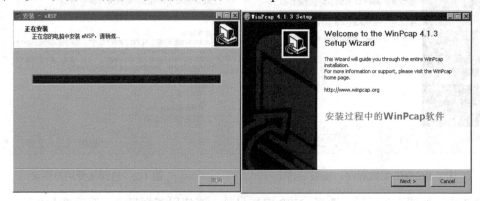

图 1-7　安装附属软件 WinPcap

第 8 步，如图 1-8 所示，安装报文分析软件 Wireshark，然后单击"Next"按钮。

第 9 步，如图 1-9 所示，选中 Wireshark 软件中的所有选项和工具，然后单击"Next"按钮。

 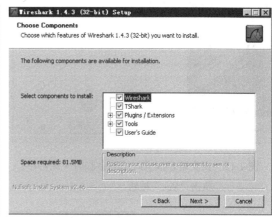

图 1-8　安装报文分析软件 Wireshark　　　图 1-9　选中 Wireshark 软件中的所有选项和工具

第 10 步，如图 1-10 所示，安装虚拟机软件 VirtualBox，然后单击"Next"按钮。

第 11 步，如图 1-11 所示，将虚拟机安装到英文目录下，然后单击"Next"按钮。

 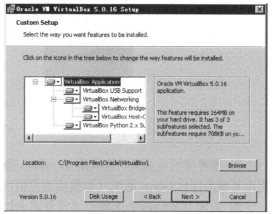

图 1-10　安装虚拟机软件 VirtualBox　　　图 1-11　将虚拟机安装到英文目录下

第 12 步，如图 1-12 所示，在安装虚拟机时可能会临时中断网络，单击"Yes"按钮。

第 13 步，如图 1-13 所示，安装一系列虚拟机插件，请全部信任。

第 14 步，如图 1-14 所示，eNSP 安装成功。

eNSP 存在一些安装兼容性问题，在 eNSP 常见安装问题中错误 40 最为常见，它将影响读者使用，解决方案通常包括以下步骤。

① 重新安装 / 升级 VirtualBox 软件（推荐 5.1.24/5.1.26 版本）。

图 1-12　在安装虚拟机时可能会临时中断网络，单击"Yes"按钮

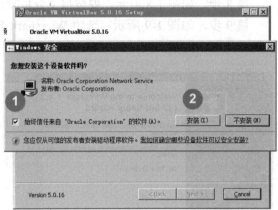
图 1-13　安装一系列虚拟机插件，请全部信任

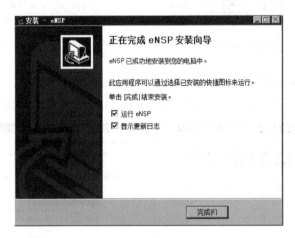
图 1-14　eNSP 安装成功

② 推荐使用专业版、旗舰版 Windows 操作系统，不推荐使用教育版、家庭版操作系统，更不要使用任何 Ghost 系统（如果遇到该情况，极有可能需要安装纯净版操作系统）。

③ 关闭 Windows 防火墙或者设置防火墙，使其允许 eNSP 进程。

④ 不推荐使用任何系统管家软件或者清理垃圾软件。

⑤ 更多内容请参考网址：https://blog.51cto.com/enderjoe/2361981。

在安装完毕 eNSP 之后，可以直接导入笔者为大家准备好的实施拓扑，即 eNSP 学习整体拓扑，如图 1-15 所示。图 1-15 中各数字所标示按钮的作用如下：❶处为开启设备，也可以选中部分设备启动；❷处为关闭所有设备，请注意是否需要在关闭设备之前保存配置，保存配置需要在配置界面的用户模式下键入 save；❸处为保存 eNSP 拓扑，请注意在配置界面键入 save 之后，单击❸处进行保存后再单击❷处关闭设备；❹处为关闭 eNSP，请注意在关闭时，还会提示是否保存，请再次单击"是"按钮保存，即总共 3 次保存才能确保配置保存成功，保存 eNSP 拓扑如图 1-16 所示。在后面学习华为网络时都会使用该拓扑。

案例 1　模拟器的部署和连接管理

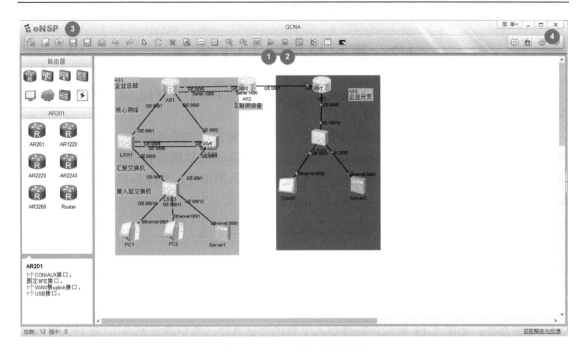

图 1-15　eNSP 学习整体拓扑

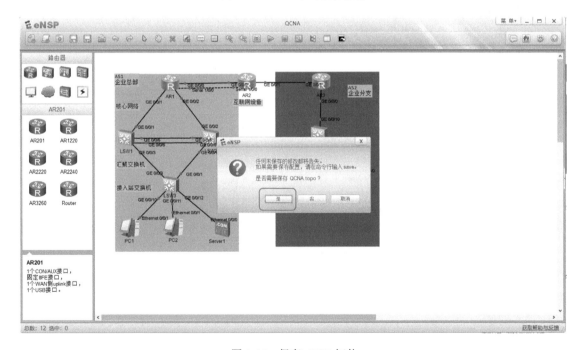

图 1-16　保存 eNSP 拓扑

那么如何操控网络设备呢？针对 eNSP 模拟器，在开启设备之后，可以双击设备管理，但是该操作存在诸多弊端，比如很多快捷键不能使用，所以建议使用 SecureCRT 来管理网络设备，该过程请参考 1.3 节内容。

1.2 思科模拟器 EVE 部署

步骤 1：请自行下载并安装 VMware Workstation 软件并重启计算机（若不重启计算机，则虚拟机软件有可能不能正常运行）。

步骤 2：使用 VMware Workstation 软件导入已经封装完毕的模拟器。

① 在安装完毕 VMware Workstation 软件之后，从本书提供的资料中下载思科模拟器，双击 QYTCNA 软件，使用 VMware Workstation 软件导入模拟器，如图 1-17 所示。

图 1-17　使用 VMware Workstation 软件导入模拟器

② 接下来将进行模拟器导入步骤。

如果必要，如图 1-18 所示，选择导入文件的存储路径（图中选择部分）。在导入过程过程中，可能会遇到"硬件合规性"检查，不需要担心，解决方案是：勾选图 1-19 中的"不再显示此消息(S)"，忽略硬件合规性检查即可。

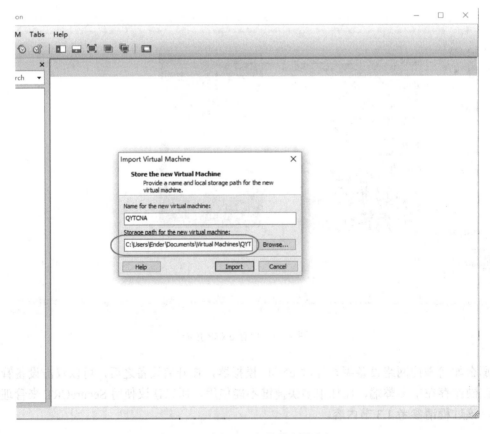

图 1-18　选择导入文件的存储路径

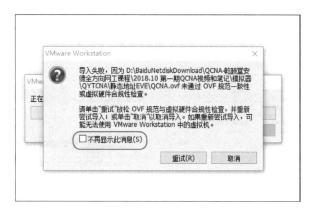

图 1-19　忽略硬件合规性检查

步骤 3：导入成功后，开始调整虚拟机的部分参数。

最重要的配置是部署网卡桥接，使个人计算机和虚拟系统通信（路由器和交换机位于虚拟系统中）。读者需要根据自己的计算机配置，参考图 1-20，部署虚拟机的重要参数。具体方法是：单击❶处，调整内存，通常调整到 3GB 足矣；之后单击❷处，通常进程数为 2 即可。

部署虚拟机网卡桥接如图 1-21 所示，此处赋予了 2*2 个核心，读者可以选择图 1-21 中❶处的自动模式或者❷处的 "Intel VT-x or AMD-V"，如果出现不支持的情况，请重启计算机进入 BIOS 开启 CPU 虚拟化功能。

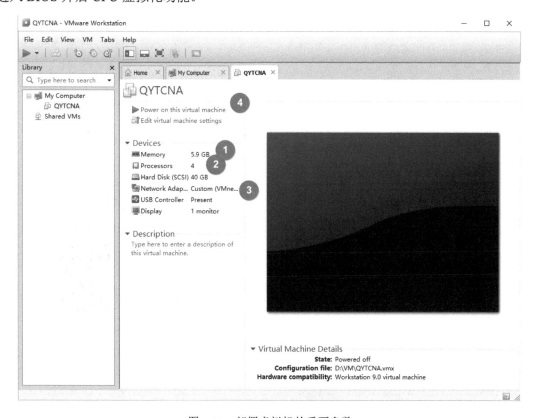

图 1-20　部署虚拟机的重要参数

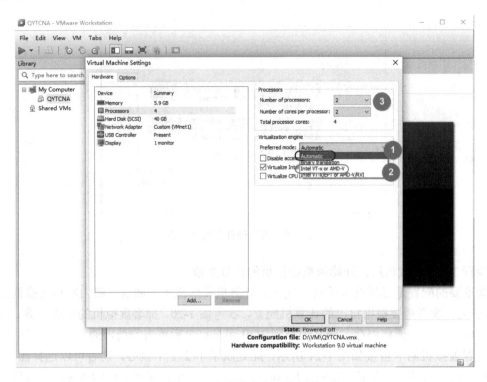

图 1-21　部署虚拟机网卡桥接

步骤 4：最重要的一步，是部署虚拟网卡，如图 1-22 所示。

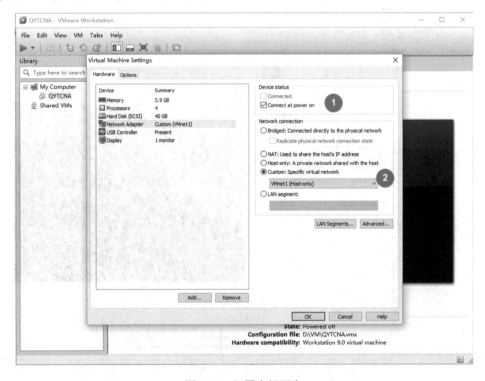

图 1-22　部署虚拟网卡

在①处选择"Connect at power on"即"开机自动连接网络"(默认已经选择);在②处,选择"Custom: Specific virtual network"即"自定义:特定网络",选择"VMnet1{Host-only}"(仅主机模式)。

步骤5:进行个人计算机的虚拟网卡设置,找到图1-23中的VM1(①处),设置VM网卡的IP地址,使得该地址和虚拟系统的网卡地址(该地址已经设为固定IP地址192.168.233.128)在同一网络,然后就可以管理虚拟网络系统了,本例中的IP地址为192.168.233.129。

图1-23 设置VM网卡的IP地址

步骤6:开始使用思科网络模拟器。

请自行安装Chrome浏览器或Firefox浏览器(其他浏览器可能出现兼容性问题)。在浏览器中输入IP地址192.168.233.128,会出现登录界面,如图1-24所示,输入用户名:admin,输入密码:qytang。

图1-24 登录界面

登录成功后，选中拓扑并打开，如图 1-25 所示。具体方法：单击图中❶处，然后选中图中❷处的 Open，思科模拟环境如图 1-26 所示。

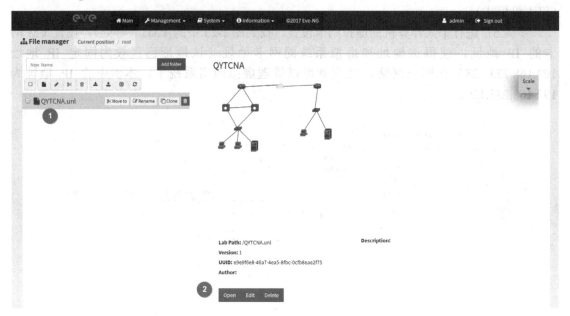

图 1-25　选中拓扑并打开

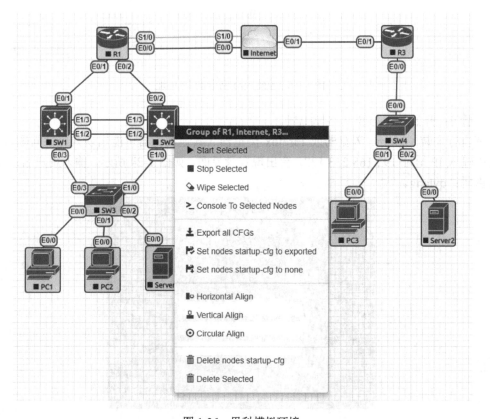

图 1-26　思科模拟环境

图 1-26 中展示了如何开启设备,读者可以选中部分或全部设备,通过鼠标右键开启或关闭设备。

在开启所有设备后,如何操控网络设备呢?需要使用 SecureCRT 来管理网络设备,具体方法请参考 1.3 节。

部署模拟器的常见问题:

① 安装 VMware Workstation 软件后务必重启计算机。

② 务必开启计算机的 CPU 支持虚拟化,如果个人的计算机没有开启,需要进入 BIOS 界面自行开启,不同 PC 的设置不同,请搜索相关内容即可。

③ 为解决浏览器兼容性问题,强烈推荐使用 Chrome 浏览器或者 Firefox 浏览器。

1.3 部署 SecureCRT 管理网络设备

SecureCRT 几乎是所有网络工程师必备的工具,所以在安装完毕 SecureCRT 之后,本节将介绍如何部署该工具来管理网络设备。

1.3.1 部署终端管理软件管理华为模拟设备

步骤 1:确定 eNSP 设备的串口号(即端口号),每个设备的串口号必须唯一。如图 1-27 所示,右键单击设备,在配置界面查看串口号。

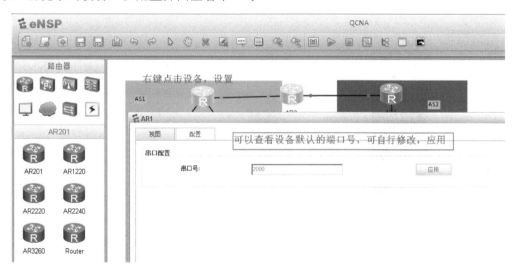

图 1-27 右键单击设备,在配置界面查看串口号

步骤 2:如图 1-28 所示,使用 SecureCRT 连接网络设备。单击❶处的"快速连接",设置❷处的 Telnet 协议,在❸处的主机位置键入 127.0.0.1(代表本机,如果在服务器上安装 eNSP,可以键入服务器地址);在❹处键入对应端口号,每个端口号对应一个设备,注意此处应该键入 2000,而图中有意错误地键入了端口号 2001,这意味着在管理另外一个设备。强烈建议读者修改每台设备的主机名,否则这意味着读者还是一名"小白",设想一下,如果所有设备都叫作 huawei,那要怎么样进行实战练习或排除故障呢?

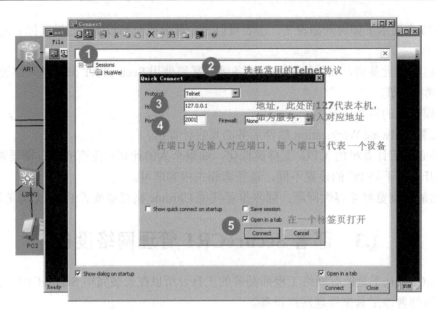

图 1-28　使用 SecureCRT 连接网络设备

步骤 3：解决 SecureCRT 兼容 eNSP 时键入命令不完整的问题。

如图 1-29 所示，在 SecureCRT 中键入命令不完整解决方案。在成功连接 eNSP 之后，键入命令"sy"然后尝试按 Tab 键补全命令，会发现出现图中标识错误，这将严重影响配置。

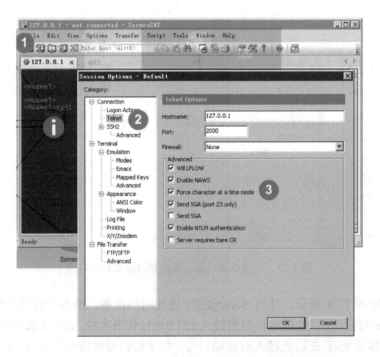

图 1-29　在 SecureCRT 中键入命令不完整解决方案

用鼠标右键单击图 1-29 中的标签，断开连接，然后选中会话，选择❷处的 Telnet 协议，之后选中❸处的"Force character at a time made"(强制每次一个字符)即可。请再次按回车键使用模拟器,即可解决该问题,其他设备请自行连接。

1.3.2 部署终端管理软件管理思科模拟设备

步骤 1:查看网络设备的 IP 地址和端口号,如图 1-30 所示。

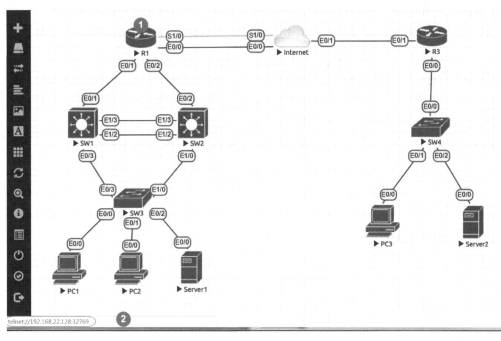

图 1-30 查看网络设备的 IP 地址和端口号

在开启 EVE 设备之后,请参照图 1-30,将鼠标置于设备之上,如图中❶处;参照图中❷处,可以看到页面左下角会出现 IP 地址(本例中的 IP 地址,和读者实际操作时所显示的 IP 地址可能并不相同)和对应的端口号。

步骤 2:使用 SecureCRT 连接思科网络设备,如图 1-31 所示。

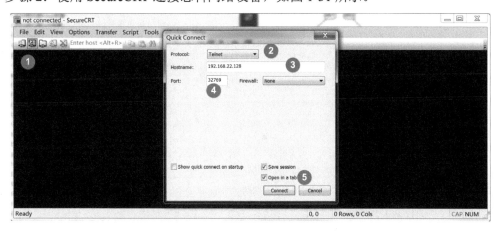

图 1-31 使用 SecureCRT 连接思科网络设备

依次单击图 1-31 中的❶处的"快速连接",在❷处选择"Telnet"协议,在❸处键入 IP 地址(即读者在图 1-30 中所见的 IP 地址),在❹处端口号位置键入对应的端口号,本例为 32769。按下回车键就可以使用网络设备了,其他设备请自行连接(只是设备端口号不同而已)。现在读者就可以通过 CLI(命令行模式)学习网络技术了。

在本案例的最后,再介绍一下 SecureCRT 常用快捷组合键,如表 1-1 所示,以方便读者加快配置速度(配置速度也是验证读者是否是一个成功的网络工程师的标准之一,不夸张地说,当您把手放到键盘上的那一刻,明眼人就已经知道您是否是合格的网络工程师了)。

表 1-1 SecureCRT 常用快捷组合键

组合键	功能
Ctrl+Tab	下一个标签
Ctrl+Shift+Tab	上一个标签
Alt+数字键	快速切换到数字的标签(不能超过 9)
Ctrl+F4	快速关闭当前标签
Ctrl+C(或者 Insert)	复制当前选中内容
Shift+Insert(或者直接单击鼠标右键)	粘贴当前选中内容

至此,本案例实施完毕。

案例 2　网络设备初始化及 Console 端口密码认证

Console 端口本地管理简书

Console 端口用于在"正经网络设备"上进行初始化管理，通常情况下，家用设备和一些低端设备不带 Console 端口管理功能，而只有 Web 管理功能。在后台管理终端上通过超级终端（通常端口 Windows 系统已不自带）、SecureCRT、Xshell、PuTTY 等类似工具对网络设备进行操作与维护。在设备初始化时，大部分专业设备上不具备 IP 地址，这意味着不能实施远程管理，此时需要通过 Console 端口来进行初始化和基本配置。Console 端口为 RJ45 接口，与后台管理终端的 COM 端口之间通过串行线缆连接，连接线缆一侧使用 RJ45 接口，另外一侧使用 DB9 接口。图 2-1 是一根实际的 Console 端口线缆。使用 Console 端口线缆来连接交换机或路由器的 Console 端口与计算机的 COM 端口，这样就可以通过计算机实现本地调试和维护。

图 2-1　一根实际的 Console 端口线缆

大部分笔记本式计算机不提供 COM 端口，所以很多时候需要 RS232 转接线缆（随着技术进步，现在出现了一侧为 Console 端口另一侧直接为 USB 接口的管理线缆；更方便的工具是带蓝牙功能的管理设备，读者可以在相关线上平台搜索），这种线缆一侧接入 Console 端口（初学者往往由于 Console 端口和业务端口外观相同，而把 Console 端口线缆错误地插入业务端口，真实设备的 Console 端口如图 2-2 所示），另一侧接入计算机的 USB 接口，在安装对应驱动软件后方可使用。为了本地管理网络设备，读者需要准备的设备或驱动软件包括：

① Console 端口线缆。
② 网络设备。
③ RS232 转接头和驱动。

④ SecureCRT（Windows 已经不再提供超级终端）等管理软件。

图 2-2　真实设备的 Console 端口

如何进行 Console 端口管理呢？读者还需要在终端管理软件中正确地进行设置。如图 2-3 所示，在 Windows 系统找到对应的 COM 端口，图中❷处为 COM3 端口。如图 2-4 所示，在终端管理软件中使用 COM 端口进行本地管理，请使用 Serial 协议，找到❸处对应的 COM 端口，在❹处波特率选择"9600"（大部分大厂商会使用该波特率，少数小厂商会使用其他波特率，如"115200"等，请注意阅读对应厂商手册），在❺处不要选择任何选项。按回车键即可进行管理，这些管理与我们使用的命令行实施是一个界面，此处省略。

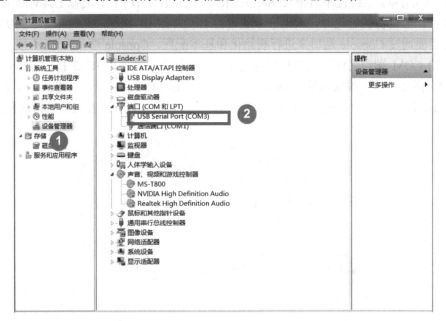

图 2-3　在 Windows 系统找到对应的 COM 端口

网络设备的管理方式一般包括以下几种。
① Console 端口（有些设备使用 Mini USB）本地管理，通过 Console 端口进行本地访问（权限较高，完成初始化配置，本地命令行实施）完成管理。
② 专有的以太管理接口 MGT（带内管理和带外管理）。
③ 远程管理：Telnet/SSH，使用 Telnet 终端访问（明文管理，方便快捷，安全性较差，即 VTY）；使用 SSH 终端访问（密文管理，安全性很好，速度较快）。
④ 跳转方式，即通过堡垒机或中间设备访问远程设备。
⑤ 通过 AUX 端口远程访问（应用较少），一些厂商设备的 AUX 端口就是 Console 端口。
可能读者没有真实设备来学习，没有关系，模拟器是学习的利器。接下来就使用模拟器

进行 Console 端口的配置学习。请读者自行启动 EVE 模拟器和 eNSP 模拟器中的设备。

图 2-4　在终端管理软件中使用 COM 端口进行本地管理

2.1　思科设备初始化及 Console 端口密码认证

启动完毕思科设备之后，读者会在 SecureCRT 中看到如下界面，那么怎么进入思科的网络设备呢？

```
--- System Configuration Dialog ---
Would you like to enter the initial configuration dialog? [yes/no]:
% Please answer 'yes' or 'no'.    //需要进入初始化会话吗？这些初始化配置一般包括名称、管理 IP
地址等，对于专业工程师来说这些基本内容都可以在 CLI 命令行中实施
Basic management setup configures only enough connectivity
for management of the system, extended setup will ask you
to configure each interface on the system

Would you like to enter basic management setup? [yes/no]: yes    //此处我们模拟错误地键入了"yes"，
即进入初始化配置，方便读者应对这种问题
Configuring global parameters:

    Enter host name [Router]: Test    //对话框中需要输入设备名，此处键入的设备名为"Test"，之后
退出初始化配置，请键入 CTRL+C 中断初始化配置
```

当通过 Console 端口进行本地管理时，需要特别注意 Console 端口的安全问题，所以现网中都会对 Console 端口开启登录认证。

Console 端口认证实施和验证操作如下。

```
按下回车键
Router>enable    //从默认的用户模式进入特权模式，特权模式是进行其他配置的前提
Router#configure te    //进入配置模式，此处使用了部分命令，请不要把命令的所有字母打全，这
才是网络工程师需要做的
Router#configure terminal    //使用 Tab 键补全命令
```

```
Enter configuration commands, one per line.    End with CNTL/Z.
Router(config)#line console ?
  <0-0>    First Line number

Router(config)#line console 0          //通常每个设备只有一个 Console 管理端口
Router(config-line)#password qytang    //设置 Console 端口的登录密码为 qytang
Router(config-line)#login              //开启登录认证，否则不会进行登录的认证
Router(config-line)#exit               //退出
Router(config)#exit
Router#
*Oct 31 21:27:25.763: %SYS-5-CONFIG_I: Configured from console by console
Router#exit        //从特权模式退出到用户模式，然后退出 Console 端口
```

测试 Console 端口登录认证：

```
Router con0 is now available      //此时通过 Console 端口方式登录
Press RETURN to get started.      //按下回车键
User Access Verification

Password:
Password:    //键入密码 qytang，按下回车键
Router>
Router#show users    //通过该验证命令查看登录的用户
    Line         User        Host(s)              Idle          Location
*   0 con 0                  idle                 00:00:00
    //*代表当前登录的线缆，读者可以看到此处为 Console0

  Interface    User        Mode                 Idle          Peer Address
```

至此，Console 端口登录方式实施成功，在测试完毕后，请读者自行去掉认证，方便后续实验。

2.2 华为设备初始化及 Console 端口密码认证

华为设备在初始化方面与思科设备非常类似，但大部分真机需要使用默认的密码认证后才能进入设备系统，读者可以参阅产品说明书，或者查询以下链接：http://support.huawei.com/onlinetoolsweb/pqt/index.jsp，这是华为的默认账号 / 密码查询工具，如图 2-5 所示。在用户键入对应设备型号之后会显示默认的用户名和密码。

当用户使用 eNSP 时，设备启动完毕后，可以直接按回车键进入系统。所以可以直接进行 Console 端口的登录认证，读者可以使用 eNSP 中任意一台设备来完成。

```
The device is running!

< Huawei >           //华为设备默认进入用户视图，用< >表示
< Huawei > system-view          //进入系统视图，进入系统视图后才可以配置其他内容
[Huawei]user-interface console 0      //进入 Console 端口的管理模式
[Huawei -ui-console0]set authentication password cipher qytang    //设置 Console 端口登录密码，cipher
表示以密文方式显示口令，而非"一目了然"的方式。请牢记密码。华为设备上命令的一致性在此并不完整，
```

可能读者运行的命令稍有不同

    [Huawei -ui-console0]quit   //退出当前模式
    [Huawei]quit
    < Huawei >quit   //从 Console 端口用户模式退出以便于测试

        Configuration console exit, please press any key to log on   //已经退出 Console 端口用户模式,请按下任意键登录,读者可按回车键
    Password: //测试,请键入密码 qytang
    < Huawei >   //成功登录
    <Huawei>display users   //验证命令,查看登录的用户情况
      User-Intf   Delay   Type   Network Address   AuthenStatus   AuthorcmdFlag
    + 0   CON 0   00:00:00   pass
       Username : Unspecified   //当前已经通过 Console 端口方式登录,并未指定用户名

    <Huawei>

图 2-5　华为的默认账号 / 密码查询工具

至此，Console 端口登录方式实施成功，在测试完毕后，请自行去掉认证，方便后续实验，针对华为设备，后续还会继续介绍以 AAA 方式登录设备。

2.3 华三设备初始化及 Console 端口密码认证

下面是在华三设备上设置 Console 端口密码认证的案例，读者可以使用任意单台设备实施。

```
<H3C>Automatic configuration is running, press CTRL_C or CTRL_D to break.   //设备自动配置正在运行，请按 Ctrl+C 键或者 Ctrl+D 键打断
<H3C> system-view    //进入系统视图
[H3C] user-interface con 0    //进入 Console 端口用户界面视图
[H3C-ui-console0] authentication-mode password    //设置通过 Console 端口登录交换机的用户进行 Password 认证
[H3C-ui-console 0] set authentication password simple qytang    //设置认证密码方式为简单即可视的密码，密码为 qytang，当然由于华三模拟器 HCL 的问题，用户查看到的配置可能是如下内容。
[H3C-line-console0]dis th
#
line aux 0
 user-role network-operator
#
line con 0
 authentication-mode password
 user-role network-admin
 set authentication password hash h$6$a9OXdEKl3AOKkloN$ZQtZzkppODJaKMm+ZFOzZAXfOfrV/4mUujl6RUYK5f6k1GaHknfPj8I+OI5Fg4AMrNHMaL/0hy05+lOQNw3Irw==
```

测试如下：

```
[R1]quit    //退出系统视图
<R1>quit    //退出 Console 端口用户视图
******************************************************************
* Copyright (c) 2004-2017 New H3C Technologies Co., Ltd. All rights reserved.*
* Without the owner's prior written consent,                      *
* no decompiling or reverse-engineering shall be allowed.         *
******************************************************************


Line con0 is available.



Press ENTER to get started.    //按回车键进入系统
Password:    //键入密码 qytang
<R1>%Dec 15 14:13:23:774 2018 R1 SHELL/5/SHELL_LOGIN: Console logged in from con0.
<R1>display users    //验证登录的用户
  Idx    Line    Idle    Time        Pid    Type
```

```
  + 0         CON 0      00:00:01    Dec 15 14:13:23       209    //当前用户通过 Console 端口登录成功

    +     : Current operation user.
    F     : Current operation user works in async mode.
```

注意：在一些华三设备上可能找不到 Console 端口命令，此时请使用 AUX 端口进行配置，AUX 端口即 Console 端口。

至此，Console 端口登录方式实施成功，在测试完毕后，请读者自行去掉认证，方便后续实验。

案例 3 配置明文远程管理协议

远程管理简书

远程管理,即通过网络方式去管理计算机设备,而不是面对面的管理计算机设备。常见的远程管理协议包含 Telnet 协议和 SSH 协议,Telnet 协议属于明文管理方式,SSH 协议属于安全的密文管理方式。

Telnet 协议在 TCP/IP 协议族中属于应用层协议,通过网络提供远程登录和虚拟终端功能。以服务器／客户端(Server/Client)模式工作,Telnet 客户端向 Telnet 服务器发起请求,Telnet 服务器提供 Telnet 服务。设备支持 Telnet 客户端和 Telnet 服务器功能,其应用端口号为 TCP 的 23 端口,常用于安全性一般的网络管理(它是一种明文传输的协议)。

3.1 在思科设备上配置远程管理协议

3.1.1 通过 Telnet 协议远程管理思科网络设备

在本案例中,要求网络工程师如图 3-1 所示,通过 Telnet 协议远程登录并管理网络设备,作为客户端的 R1 使用 Telnet 协议,远程管理服务器 SW1。

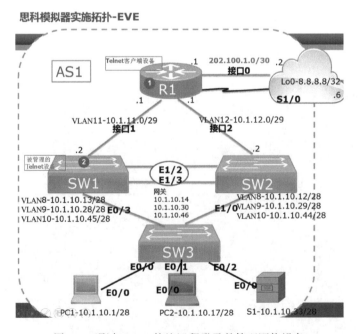

图 3-1 通过 Telnet 协议远程登录并管理网络设备

3.1.2 配置 Telnet 协议拓扑说明

在图 3-1 中，❶处即 R1 的 IP 地址为 10.1.11.1/29，R1 作为 Telnet 协议的客户端；❷处即 SW1 的 IP 地址为 10.1.11.2/29，其作为 Telnet 的服务器。注意 SW1 请使用默认的 VLAN1。

3.1.3 远程管理配置要点

① 需要给交换机配置一个管理 IP 地址，如果 PC 与交换机不在同一个网段，则需要给交换机配置默认网关，实施路由协议，以确保设备之间可以通信。

② 需要配置一个 enable 密码及 Telnet 密码才能完全实现远程登录思科网络设备。

3.1.4 协议配置步骤

① 请先配置设备互连的 IP 地址以完成设备通信。

```
Router(config)#hostname R1    //修改设备 1 的名字，从默认的 Router 修改为 R1，请读者养成良好习惯，修改每个设备的名字方便对后续故障进行排错
R1(config)#int e0/1    //进入设备的 e0/1 接口，初学者往往出现键入问题，请注意
R1(config-if)#no shu    //思科的路由器接口默认关闭，配置开启接口命令
R1(config-if)#no shutdown
R1(config-if)#ip address 10.1.11.1 255.255.255.248    //配置接口地址，/29 等于 255.255.255.248
R1(config-if)#do sh ip int brief    //思科设备在非特权模式键入 show 命令时需要加 do 命令，华为和华三设备不需要该方式

Interface         IP-Address      OK? Method Status                 Protocol
Ethernet0/0       unassigned      YES unset  administratively down  down
Ethernet0/1       10.1.11.1       YES manual up                     up
!
SW1(config)#interface vlan 1    //进入交换机 1 的管理 VLAN 接口
SW1(config-if)#no shutdown    //交换机的 SW1（交换机的 VLAN 接口）默认关闭，不要忽略开启
SW1(config-if)#
*Nov  2 20:52:58.177: %LINK-3-UPDOWN: Interface VLAN1, changed state to up
*Nov  2 20:52:59.182: %LINEPROTO-5-UPDOWN: Line protocol on Interface Vlan1, changed state to up
SW1(config-if)#ip address 10.1.11.2 255.255.255.248    //赋予 IP 地址
SW1#show ip int brief
Interface         IP-Address      OK? Method Status      Protocol
Ethernet0/0       unassigned      YES unset  up          up
Ethernet0/1       unassigned      YES unset  up          up
Ethernet0/2       unassigned      YES unset  up          up
Ethernet0/3       unassigned      YES unset  up          up
Ethernet1/0       unassigned      YES unset  up          up
Ethernet1/1       unassigned      YES unset  up          up
Ethernet1/2       unassigned      YES unset  up          up
Ethernet1/3       unassigned      YES unset  up          up
Vlan1             10.1.11.2       YES manual up          up
SW1#ping 10.1.11.1    //设备之间必须能够进行网络通信，才可以进行远程管理
Type escape sequence to abort.
Sending 5, 100-byte ICMP Echos to 10.1.11.1, timeout is 2 seconds:
```

!!!!!
Success rate is 100 percent (5/5), round-trip min/avg/max = 1/1/5 ms

② 在完成基本的网络通信之后，配置交换机的远程登录 VTY（Virtual Teletype Terminal，虚拟终端），即开启 Telnet 服务。

```
SW1(config)#line vty 0         //开启 VTY 的通道 0，具体定义的通道数需要自行决定
SW1(config-line)#password qytang    //设置进入该虚拟通道的密码
SW1(config-line)#transport input telnet   //允许 Telnet 协议来管理本设备，某些设备的系统有可能默认不允许任何协议管理
```

在配置完远程登录后，进行测试和验证：

```
R1#telnet 10.1.11.2
Trying 10.1.11.2 ... Open


User Access Verification

Password:    //输入密码，注意密码并不显示

SW1>enable
% No password set    //无法进入设备的特权模式，是因为没有设置进入特权模式的密码。该密码类似家中防盗门的密码
```

补充 SW1 的密码：

```
SW1(config)#enable password qytang
SW1>enable
Password:    //请键入密码 qytang
SW1#
SW1#show users
    Line       User     Host(s)       Idle         Location
*   0 con 0             idle          00:00:00
    2 vty 0             idle          00:01:41 10.1.11.1    //读者可以看到位于 10.1.11.1 的用户通过 vty0 通道登录了本设备（SW1）

    Interface  User     Mode          Idle         Peer Address
```

至此，基本的思科设备的 Telnet 远程管理实施和验证完毕。

3.2 在华为设备上配置远程管理协议

3.2.1 通过 Telnet 协议远程管理华为网络设备

通过 Telnet 功能远程登录管理华为设备如图 3-2 所示。作为客户端的 R1 可以通过网络使用 Telnet 远程管理服务器 SW1，以便后期在 SW1 上实施网络配置、修改配置和测试等工作。

3.2.2 配置远程管理协议拓扑说明

如图 3-2 所示，R1 和 SW1 的 IP 地址分别为 10.1.11.1/29 和 10.1.11.2/29，注意在 SW1 上请使用默认的 VLAN1 作为管理 VLAN，关于 VLAN 会在后续内容中讲解。

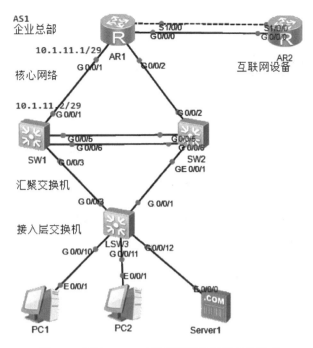

图 3-2 通过 Telnet 功能远程登录管理华为设备

3.2.3 远程管理配置要点

① 需要给交换机配置一个管理 IP 地址，如果 PC 与交换机不在同一个网段，则需要给交换机配置一个默认网关，确保设备之间可以通信。

② 需要配置登录设备的级别，否则无法进入远程设备。

3.2.4 远程管理配置详解

① 完成网络设备基本通信功能的相关配置。

```
<Huawei >system-view
[Huawei]sysname HW-R1       //修改 R1 的名字，以便识别
[HW-R1]int g0/0/1           //进入设备的 g0/0/1 接口
[HW-R1-GigabitEthernet0/0/1]ip address 10.1.11.1 29   //配置接口的 IP 地址，华为设备上可以直接键入/29 长度的掩码
!
[Huawei]sysname SW1
[SW1]interface vlan 1       //进入 SW1 设备的管理接口，即 VLAN1 接口
[SW1-Vlanif1]ip address 10.1.11.2 29   //配置管理 VLAN（管理接口）。华为设备不同于思科设备，在物理上是默认开启的，赋予 IP 地址后可以工作，不要忘记掩码（29）
[SW1-Vlanif1]ping 10.1.11.1  //测试网络连接，不需要加 do 命令。此时设备可以通信
  ping 10.1.11.1: 56   data bytes, press CTRL_C to break
    Reply from 10.1.11.1: bytes=56 Sequence=1 ttl=255 time=370 ms
    Reply from 10.1.11.1: bytes=56 Sequence=2 ttl=255 time=60 ms
    Reply from 10.1.11.1: bytes=56 Sequence=3 ttl=255 time=60 ms
    Reply from 10.1.11.1: bytes=56 Sequence=4 ttl=255 time=40 ms
```

```
        Reply from 10.1.11.1: bytes=56 Sequence=5 ttl=255 time=60 ms

        --- 10.1.11.1 ping statistics ---
          5 packet(s) transmitted
          5 packet(s) received
          0.00% packet loss
          round-trip min/avg/max = 40/118/370 ms
```

② 在前置条件实现之后，请开启设备的 Telnet 服务功能。

```
[SW1-Vlanif1]quit
[SW1]telnet server enable         //开启 Telnet 服务功能
[SW1]user-interface vty 0 4       //进入 VTY 的 0～4 通道
[SW1-ui-vty0-4]set authentication password simple qytang   //开启远程通道0～4，设置简单认证密码 qytang
```

③ 实施远程登录测试。

```
<R1>telnet 10.1.11.2       //注意，测试的模式为用户模式
  Press CTRL_] to quit telnet mode
  Trying 10.1.11.2 ...
  Connected to 10.1.11.2 ...

Login authentication


Password:      //键入密码 qytang
Info: The max number of VTY users is 5, and the number
      of current VTY users on line is 1.
      The current login time is 2018-11-02 21:14:37.
<SW1>sy     //无法进入远程设备的系统视图，无法补全配置命令，原因是用户登录级别不够
[SW1-ui-vty0-4]user privilege level 15    //用户远程登录的级别赋予最高级别 15 级
```

④ 再次测试。

```
<HW-R1>telnet 10.1.11.2
  Press CTRL_] to quit telnet mode
  Trying 10.1.11.2 ...
  Connected to 10.1.11.2 ...

Login authentication


Password:
Info: The max number of VTY users is 5, and the number
      of current VTY users on line is 1.
      The current login time is 2018-12-15 15:14:03.

<HW-SW1>sy    //使用 Tab 键补全命令
<HW-SW1>system-view
Enter system view, return user view with Ctrl+Z.
[HW-SW1]display users
```

User-Intf	Delay	Type	Network Address	AuthenStatus	AuthorcmdFlag
0 CON 0	00:00:33				no
Username : Unspecified					
+ 34 VTY 0	00:00:00	TEL	10.1.11.1	pass	no
Username : Unspecified	//位于 10.1.11.1 的客户端通过 Telnet 协议登录了 SW1 的 VTY 通道 0				

至此，华为设备的 Telnet 服务配置实施完毕。

3.3 在华三设备上配置远程管理协议

3.3.1 通过 Telnet 协议远程管理华三网络设备

通过 Telnet 功能远程登录管理华三设备如图 3-3 所示。作为客户端的 R1 可以使用 Telnet 协议远程管理服务器 SW1。

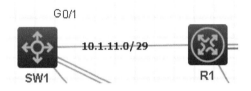

图 3-3 通过 Telnet 功能远程登录管理华三设备

3.3.2 华三设备远程管理协议配置拓扑

在本例中，把网络设备既作为 Telnet 协议的服务器端，又作为 Telnet 协议的客户端，即从一台网络设备去远程管理另外一台网络设备。在现实网络中，客户端是运行 SecureCRT 等终端的管理员计算机，而服务器是被管理的网络设备。

3.3.3 远程管理配置要点

① 需要给交换机配置一个管理 IP 地址，如果 PC 与交换机不在同一个网段，则需要给交换机配置一个默认网关，确保设备之间可以通信。

② 需要配置登录设备的级别，另外，需要开启 Telnet 服务。

3.3.4 远程管理配置详解

① 配置设备之间通信，这是一切网络应用的前提。

```
R1:
interface GigabitEthernet0/1
ip address 10.1.11.1 255.255.255.248
SW1:
interface Vlan-interface1
 ip address 10.1.11.2 255.255.255.248
[R1]ping 10.1.11.2
ping 10.1.11.2 (10.1.11.2): 56 data bytes, press CTRL_C to break
```

```
56 bytes from 10.1.11.2: icmp_seq=0 ttl=255 time=20.000 ms
56 bytes from 10.1.11.2: icmp_seq=1 ttl=255 time=3.000 ms
56 bytes from 10.1.11.2: icmp_seq=2 ttl=255 time=2.000 ms
56 bytes from 10.1.11.2: icmp_seq=3 ttl=255 time=2.000 ms
56 bytes from 10.1.11.2: icmp_seq=4 ttl=255 time=2.000 ms

--- ping statistics for 10.1.11.2 ---
5 packet(s) transmitted, 5 packet(s) received, 0.0% packet loss
round-trip min/avg/max/std-dev = 2.000/5.800/20.000/7.111 ms
!
```

② 配置交换机，开启 Telnet 服务。

```
[SW1]telnet server enable           //手动配置开启 Telnet 服务
[SW1]line vty 0 4                   //进入 VTY 通道，该命令与思科相同
[SW1-line-vty0-4]set authentication password simple qytang       //设置通过 VTY 登录的密码 qytang
[SW1-line-vty0-4]dis th
#
line aux 0
 user-role network-operator
#
line con 0
 user-role network-admin
#
line vty 0 4
 user-role network-operator       //HCL 模拟器上自带的登录用户为网络管理员，即已经通过角色授权，类比华为的用户级别
 set authentication password hash $h$6$KYi+uCCUxGF7i4NV$TQWRookb61HO/fW7I25rsttmsZOl
mm2bdodHUQ6AgMgr3gJ8HHB22WUgBqlj60YHhteeKU6YJQkobBKg9r0l9Q==      //在模拟器上配置的密码并不会明文显示
```

③ 进行远程登录测试。

```
<R1>telnet 10.1.11.2     //注意测试模式为用户模式
Trying 10.1.11.2 ...
Press CTRL+K to abort
Connected to 10.1.11.2 ...

**********************************************************************
* Copyright (c) 2004-2017 New H3C Technologies Co., Ltd. All rights reserved.*
* Without the owner's prior written consent,                          *
* no decompiling or reverse-engineering shall be allowed.             *
**********************************************************************

Password:      //请键入密码，密码并不显示
<SW1>system-view    //进入 SW1 的系统视图
System View: return to User View with Ctrl+Z.
[SW1]display users
     Idx   Line       Idle       Time              Pid    Type
     0     CON 0      00:08:13   Dec 15 15:21:21   214
```

```
  +84        VTY 0     00:00:00  Dec 15 15:40:16       267         TEL  //远程用户已经成功通过
VTY0 登录 SW1

       Following are more details.
       VTY 0    :
                Location: 10.1.11.1
          +     : Current operation user.
          F     : Current operation user works in async mode.
```

至此，远程登录实施完毕。

如果出现远程登录故障，请参考如图 3-4 所示的 Telnet 登录排错思路。

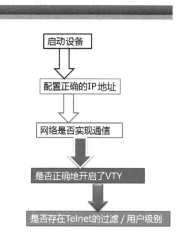

图 3-4　Telnet 登录排错思路

案例 4　配置 SSH 协议

SSH 协议安全远程管理简书

Telnet 传输过程采用 TCP 进行明文传输，缺少安全的认证方式，容易招致 DoS（Denial of Service，拒绝服务）、主机 IP 地址欺骗和路由欺骗等恶意攻击，存在很大的安全隐患。

相对于 Telnet，STelnet 基于 SSH2 协议，客户端和服务器端之间经过协商，建立安全连接，客户端可以像操作 Telnet 一样登录服务器端。SSH 协议通过以下措施实现在不安全的网络上提供安全的远程登录操作：支持 RSA（Revest-Shamir-Adleman）和 ECC（Elliptic Curves Cryptography）认证方式。客户端需要创建一对密钥（公用密钥和私用密钥），并把公用密钥发送到需要登录的服务器上。服务器使用预先配置的该客户端的公用密钥，与报文中携带的客户端公用密钥进行比较。如果两个公用密钥不一致，则服务器断开与客户端的连接。如果两个公用密钥一致，客户端继续使用自己本地密钥对的私用密钥部分，对特定报文进行摘要运算，将所得的结果（即数字签名）发送给服务器，向服务器证明自己的身份。服务器使用预先配置的该客户端的公用密钥，对客户端发送过来的数字签名进行验证。支持用加密算法 DES（Data Encryption Standard）、3DES、AES128（Advanced Encryption Standard 128）、AES256 对用户名、密码，以及传输数据进行加密。

设备支持 SSH 服务器功能，可以与多个 SSH 客户端建立连接。同时，设备还支持 SSH 客户端功能，可以与支持 SSH 服务器功能的设备建立 SSH 连接，从而实现本地设备通过 SSH 方式登录到远程设备。目前，当设备作为 SSH 服务器端时，支持 SSH2 和 SSH1 两个版本。

4.1　在思科设备上配置 SSH 协议

4.1.1　客户安全管理需求

使用 R2 作为 SSH 客户端，通过远程方式可以访问 SSH 服务器 R1，从而满足网络远程维护实施的安全要求。SSH 协议作为安全的网络管理协议，应该普遍应用到安全性较高的网络中，而不推荐使用 Telnet 协议这种明文管理协议。

4.1.2　使用 SSH 协议组网拓扑

SSH 协议配置拓扑如图 4-1 所示。图中 R2 的 IP 地址为 202.100.1.2/30，通过 SSH 协议访问登录并管理 SSH 服务器 R1，R1 的 IP 地址为 202.100.1.1/30。

案例 4 配置 SSH 协议

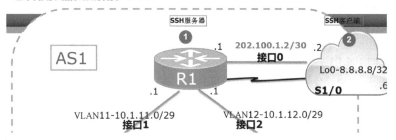

图 4-1 SSH 协议配置拓扑

4.1.3 配置 SSH 协议要点

① 需要管理员开启 SSH 功能。
② 需要手工生成 Key（密钥）。
③ 若 PC 与交换机不在同一个网段，则需要配置交换机的默认网关和路由。

4.1.4 配置 SSH 协议步骤详解

① 请先完成最基本的设备通信功能配置，这是网络管理的前提，而初学者往往会忽略这一点。

```
R1 #sh run int e0/0
interface Ethernet0/0
 no shutdown
 ip address 202.100.1.1 255.255.255.252
R2-internet#sh run int e0/0
interface Ethernet0/0
 no shutdown
            202.100.1.2 255.255.255.252
R2-internet#ping 202.100.1.1    //设备之间实现通信
Type escape sequence to abort.
Sending 5, 100-byte ICMP Echos to 202.100.1.1, timeout is 2 seconds:
.!!!!
Success rate is 80 percent (4/5), round-trip min/avg/max = 1/1/1 ms
```

② 完成 SSH 服务器的基本配置。

```
Router>enable
Router#conf t
Enter configuration commands, one per line.  End with CNTL/Z.
Router(config)#hostname R1    //配置 SSH 协议时必须修改主机名，不能使用默认设备
R1(config)#username ender password qytang    //配置登录所需要的用户名和密码
R1(config)#ip domain name qytang.com    //必须设置路由器的域名，否则无法生成 Key
R1(config)#crypto key generate rsa    //手动产生 RSA 的 Key，回车后需要键入 Key 长度，推荐 768
位以上
How many bits in the modulus [512]: 768
% Generating 768 bit RSA keys, keys will be non-exportable...
[OK] (elapsed time was 0 seconds)
R1(config)#ip ssh version 2    //开启 SSH 协议版本 2
R1(config)#line vty 0 4
```

```
    R1(config-line)#login local              //认证时采用本地认证,即使用之前在全局配置的用户名和密码
    R1(config-line)#transport input ssh      //允许以 SSH 方式登录设备
```
③ 完成 SSH 登录测试。
```
    R2-internet#ssh -l ender 202.100.1.1  //在路由器上完成 SSH 方式登录,-l 代表使用用户名 ender 登
录,目的地址为 202.100.1.1
    Password:      //请键入对应的密码:qytang

    R1>        //成功登录 R1
```
④ 验证。
```
    R1>show users
        Line       User      Host(s)           Idle         Location
    *   0 con 0              idle              00:00:00
        2 vty 0    ender     idle              00:00:16     202.100.1.2   //远程用户 ender 从远
程地址 202.100.1.2 访问了本设备

        Interface  User      Mode              Idle         Peer Address
```

以上测试是将路由器作为 SSH 的客户端进行测试的。

注意:此处使用路由器作为客户端,如果是真实设备,请采用 SecureCRT 等终端管理软件,采用 SSH 协议登录,采用 SecureCRT 实现 SSH 登录实例如图 4-2 所示,选择❶处的 SSH 协议,输入❷处的 IP 地址进行连接。

图 4-2 采用 SecureCRT 实现 SSH 登录实例

至此,思科设备的 SSH 协议配置完毕。

4.2 在华为设备上配置 SSH 协议

4.2.1 使用 SSH 协议的组网需求

通过 SSH 方式远程安全管理网络设备,用密文的方式在网络中传输管理数据,以满足企

业网络设备的安全管理需求（读者可以用模拟器抓取 Telnet 协议的报文，会发现所有的数据都可以被以明文的形式抓取到）。

4.2.2 使用 SSH 协议的组网拓扑

使用 SSH 协议的组网拓扑如图 4-3 所示，R2 作为 SSH 客户端远程登录到 SSH 服务器 R1，其 IP 地址如图所示，需要说明的一点是，华为的 eNSP 模拟器中的交换机大部分（可能）不支持 SSH 协议，所以在本例中使用了 AR2200 的设备（请勿使用 Router，它可能也不支持 SSH 协议，依据 eNSP 版本不同而有所不同）。

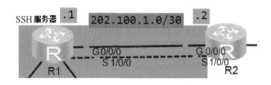

图 4-3　使用 SSH 协议的组网拓扑

4.2.3 配置 SSH 协议要点

① 需要在服务器上创建 SSH 账户并开启 SSH 协议。
② 手工创建密钥（推荐 768 位以上）。
③ VTY 的用户接口下允许开启 SSH 登录方式。

4.2.4 配置 SSH 协议步骤详解

① 请读者完成基本的网络连通性配置，同时熟悉华为设备上 IP 地址的配置过程。

```
R1:
interface GigabitEthernet0/0/0
 ip address 202.100.1.1 255.255.255.252
R2
interface GigabitEthernet0/0/0
 ip address 202.100.1.2 255.255.255.252
[R1]ping 202.100.1.2
  ping 202.100.1.2: 56   data bytes, press CTRL_C to break
    Reply from 202.100.1.2: bytes=56 Sequence=1 ttl=255 time=40 ms
    Reply from 202.100.1.2: bytes=56 Sequence=2 ttl=255 time=20 ms
    Reply from 202.100.1.2: bytes=56 Sequence=3 ttl=255 time=20 ms
    Reply from 202.100.1.2: bytes=56 Sequence=4 ttl=255 time=20 ms
    Reply from 202.100.1.2: bytes=56 Sequence=5 ttl=255 time=30 ms

  --- 202.100.1.2 ping statistics ---
    5 packet(s) transmitted
    5 packet(s) received
    0.00% packet loss
  round-trip min/avg/max = 20/26/40 ms
```

② 请在 R1 上级服务器上完成 SSH 协议配置。

[R1]aaa   //进入 AAA 模式，即认证、授权、审计模式，该模式用于创建用户名和密码
[R1-aaa]local-user ender password cipher %$%$rDTc;00VV8g2V6M,3un(a,a}%$%$
//创建本地用户 ender，密码为 qytang
[R1-aaa] local-user ender privilege level 15   //该用户的级别为最高的 15 级
[R1-aaa]local-user ender service-type ssh   //该用户用于 SSH 登录
[R1-aaa]quit   //退出 AAA 模式
[R1]ssh user ender authentication-type password   //SSH 用户 ender 通过密码进行认证
  Authentication type setted, and will be in effect next time
[R1]stelnet server enable   //开启 SSH 服务，该服务默认处于关闭状态
The STELNET server is already started.   //SSH 服务已经开启
[R1]rsa local-key-pair create   //在设备上创建 RSA 的 Key
The key name will be: Host
% RSA keys defined for Host already exist.
Confirm to replace them? (y/n)[n]:y
The range of public key size is (512 ～ 2048).
NOTES: If the key modulus is greater than 512,
       It will take a few minutes.
Input the bits in the modulus[default = 512]:768   //使用 768 位而非默认的 512 位
Generating keys...
.++++++++
..................++++++++
..................++++++++
...............++++++++

[R1]
[R1]user-interface vty 0 4   //进入 VTY 通道
[R1-ui-vty0-4]authentication-mode aaa   //在 VTY 通道的认证模式中选择 AAA 模式，即在认证时使用 AAA 定义的用户名和密码
[R1-ui-vty0-4]protocol inbound ?
   all   All protocols
   ssh   SSH protocol
   telnet Telnet protocol
[R1-ui-vty0-4]protocol inbound ssh   //VTY 允许 SSH 登录

③ 完成 SSH 登录测试。

[R2]stelnet 202.100.1.1   //注意在系统视图进行测试，202.100.1.1 为 SSH 服务器的地址
Please input the username:ender   //键入用户名 ender
Trying 202.100.1.1 ...
Press CTRL+K to abort
Connected to 202.100.1.1 ...
The server's public key does not match the one catched before.
The server is not authenticated. Continue to access it? (y/n)[n]:y   //是否继续访问 SSH 服务器，请键入 y
Dec 16 2018 11:00:50-08:00 Internet %%01SSH/4/CONTINUE_KEYEXCHANGE(l)[2]:The server had not been authenticated in the process of exchanging keys. When deciding whether to continue, the user chose Y.
[R2]
Update the server's public key now? (y/n)[n]:y   //如有必要更新服务器的公钥，键入 y。存在公钥的

原因是由于之前已经登录过

```
           Dec  16  2018  11:00:52-08:00  Internet  %%01SSH/4/UPDATE_PUBLICKEY(l)[3]:When  deciding
whether to update the key 202.100.1.1 which already existed, the user chose Y.
        [R2]
        Enter password:       //键入密码，即 qytang
        <R1>
        <R1>display users    //登录成功，查看登录的用户为 ender，位于 202.100.1.2 的设备已经通过 VTY
登录 R1
           User-Intf       Delay       Type        Network Address      AuthenStatus      AuthorcmdFlag
        + 130 VTY 1       00:00:00    SSH         202.100.1.2          pass              Username : ender


        <R1>
        <R1>system-view
        Enter system view, return user view with Ctrl+Z.
        [R1]
```

//注意：华为设备作为客户端，第一次登录可以使用命令 ssh client first-time enable

至此，华为设备的 SSH 协议配置完毕。

4.3 在华三设备上配置 SSH 协议

4.3.1 安全管理网元需求

通过 SSH 方式远程安全管理华三网络设备。正如前面所述，Telnet 协议实施方便，应用起来也非常快捷，但是它确实不那么安全，可以说所有的数据都"赤裸裸"地暴露在网络上，而 SSH 协议是一种安全传输协议，它传输的数据经过加密处理。

4.3.2 使用 SSH 协议组网拓扑

在华三设备上配置 SSH 协议拓扑如图 4-4 所示，图中 SW1 充当 SSH 客户端远程登录到 SSH 服务器 R1，地址如图所示。与前边的案例相同，在本例中，把网络设备既作为 SSH 协议的服务器端也作为 SSH 协议的客户端，即通过一台网络设备远程管理另外一台网络设备。在现实网络中，客户端是运行 SecureCRT 等终端的管理员计算机，而服务器是被管理的网络设备。

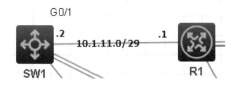

图 4-4　在华三设备上配置 SSH 协议拓扑

4.3.3 配置 SSH 协议要点

本节的配置要点与前面思科、华为设备实施 SSH 协议的配置要点几乎相同,有以下几点。

① 需要在服务器上创建 SSH 账户并开启 SSH 功能。
② 手工创建密钥。
③ VTY 允许 SSH 登录。

4.3.4 配置 SSH 协议和测试步骤

① 华三设备 SSH 服务器配置。

```
[R1]ssh server enable          //开启 SSH 服务
[R1]local-user ender           //创建本地用户,这是在华三设备上创建 AAA 用户的方式
[R1-luser-manage-ender]password simple qytang    //配置该用户的密码,注意模拟器会自动改成 hash(即加密后)的密码
[R1-luser-manage-ender]service-type ssh    //该用户服务于 SSH 应用
[R1]user-interface vty 0 4     //进入 VTY 接口
[R1-line-vty0-4]authentication-mode ?
  none       Login without authentication
  password   Password authentication
  scheme     Authentication use AAA
[R1-line-vty0-4]authentication-mode scheme   //登录 VTY 用户的认证方式为 AAA,使用前面配置的用户名和密码登录
[R1-line-vty0-4]protocol inbound ssh    //VTY 线路允许 SSH 登录管理
[R1-line-vty0-4]user-role level-15      //登录的用户为 15 级
```

② 在客户端进行 SSH 登录测试。

```
<SW1>sshv2 202.100.1.1    //注意需要在系统视图进行 SSH 登录
Please input the username:ender    //键入用户名
Trying 202.100.1.1 ...
Press CTRL+K to abort
Connected to 202.100.1.1 ...
The server's public key does not match the one catched before.
The server is not authenticated. Continue to access it? (y/n)[n]:y    //继续接入远程设备
Update the server's public key now? (y/n)[n]:y    //如有必要更新服务器端的公钥,键入 y
Enter password:    //键入密码
<R1>sy
<R1>system-view    //登录成功
Enter system view, return user view with Ctrl+Z.
```

至此,本案例实施完毕。

案例 5　配置 TFTP

TFTP（简单文件传送协议）简书

TFTP（Trivial File Transfer Protocol，简单文件传送协议）的目标是在 UDP 之上建立一个类似 FTP 但仅支持文件上传和文件下载功能的传输协议，所以它不包含 FTP 中的目录操作和用户权限等内容。

TFTP 的一个重要特点就是简单且易于实现，这也是设计 TFTP 的初衷。

优点：每个数据包大小固定，这样在进行内存分配处理的时候比较直接；实现机制简单；每个数据包都有确认机制，可靠性高。

缺点：传输效率不高；滑动窗口实现机制太简单，并且该窗口仅有一个包的大小；超时处理机制并不完善。

思科设备可以作为 TFTP 的服务器端和客户端。

华为和华三设备仅可作为 TFTP 的客户端。

5.1　在思科设备上配置 TFTP

5.1.1　使用 TFTP 的组网需求

TFTP 在思科传统 IOS 上的应用主要是从设备上复制、备份、更新 IOS 镜像文件或其他文件，在本案例中介绍通过 TFTP 服务从路由器上下载备份 IOS 文件，以及向路由器上传 IOS 文件（一个比较大的文件）。

5.1.2　使用 TFTP 的组网拓扑

思科设备 TFTP 实施拓扑如图 5-1 所示，在计算机中运行 TFTP 软件（本实验使用 Tftpd64 软件）作为 TFTP 服务器，计算机通过网络连接路由器，请自行设置计算机的 IP 地址（注意关闭 Windows 的防火墙，否则可能无法通信）。

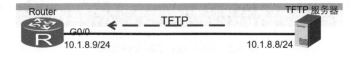

图 5-1　思科设备 TFTP 实施拓扑

5.1.3　配置 TFTP 要点

① 保证设备之间的网络可以通信。

② 开启 TFTP 软件，即用于上传或者下载文件的软件。
③ 通过命令上传或者下载 IOS 文件。

5.1.4 配置 TFTP 步骤详解

本实验无法通过模拟器实现，请使用真机操作（本例采用思科的 1921 路由器）。首先，请实现设备通信。

路由器配置：
```
R1(config)#int g0/0     //进入设备的接口，初学者往往出现键入问题，请注意
R1(config-if)#no shu    //思科的路由器接口默认关闭，配置开启接口命令
R1(config-if)#ip address 10.1.8.9 255.255.255.0   //配置路由器上连接 PC 接口的 IP 地址
PC：
C:\Users\ThinkPad>ping 10.1.8.8    //PC 和路由器通信正常
正在 ping 10.1.8.8 具有 32 字节的数据:
来自 10.1.8.8 的回复: 字节=32 时间=1ms TTL=64
来自 10.1.8.8 的回复: 字节=32 时间=1ms TTL=64
来自 10.1.8.8 的回复: 字节=32 时间=1ms TTL=64
来自 10.1.8.8 的回复: 字节=32 时间<1ms TTL=64

10.1.8.8 的 ping 统计信息:
    数据包: 已发送 = 4, 已接收 = 4, 丢失 = 0 (0% 丢失),
往返行程的估计时间(以毫秒为单位):
    最短 = 0ms, 最长 = 1ms, 平均 = 0ms
```

在服务器中运行 TFTP 软件，然后从路由器上下载文件，通过 IOS 复制文件实例如图 5-2 所示，在图中❶处开启了 TFTP 软件，在❷处键入命令以便复制设备硬盘（flash）的文件。

图 5-2　通过 IOS 复制文件实例

在图 5-2 中，❶处代表已经运行了 Tftpd64 软件，该软件会自动列出文件目录和服务器的 IP 地址，配置❷处的命令。注意，请勿自行键入所有命令，请使用 Tab 键补全命令，实现快速键入。

```
R1#copy flash:c1900-universalk9-mz.SPA.152-4.M5.bin tftp:    //从路由器上复制硬盘（即 flash）上的
```

IOS 文件（即以.bin 结尾的文件）到 TFTP 服务器，回车

图 5-2 中的地址 10.1.8.8 为 TFTP 服务器的 IP 地址，其后为目的文件名，请采用默认名即可。

另外，思科设备也可以作为 TFTP 服务器。命令如下：

  R1(config)#tftp-server flash: //思科设备也可以作为 TFTP 服务器使用，其他与该设备通信的网络设备可以从 R1 上复制文件

5.2 在华为及华三设备上实现 TFTP

5.2.1 使用 TFTP 的组网需求

使用华为或者华三设备作为 TFTP 客户端，从 TFTP 服务器上下载文件，向 TFTP 服务器上传文件，用以备份系统文件、日志、配置文件等。

5.2.2 使用 TFTP 的组网环境

如图 5-3 所示，在华为或华三设备上运行 TFTP 拓扑（本例采用 Tftpd64），作为 TFTP 服务器，和交换机（本例使用华为交换机 5700，模拟器不能模拟此实验）在同一网络中，地址如图 5-3 所示，PC 的 IP 地址为 10.1.8.8，交换机的 IP 地址为 10.1.1.9。

图 5-3 在华为或华三设备上运行 TFTP 拓扑

5.2.3 配置 TFTP 要点

① 保证 TFTP 服务器和交换机可以通信（本例中为直连网络，也可以跨越网络实现）。
② 运行 TFTP 软件。
③ 运行正确的 TFTP 客户端命令。

5.2.4 配置 TFTP 步骤详解

① 请保证网络设备可以与 TFTP 服务器通信。（注意：很多读者的计算机系统的防火墙默认开启，它极有可能导致其计算机无法被 ping 通，但此时计算机其实可以实现通信。读者可以关闭系统防火墙尝试。）

运行 TFTP 软件示意图如图 5-4 所示。

② 配置交换机 IP 地址，保证该地址和 PC 属于同一网络。配置交换机地址的命令如下。

  [SW5-HW]int vlan 1
  [SW5-HW-Vlanif1]ip address 10.1.8.9 24 //本例采用交换机默认的 VLAN1，在现网中请根据需要配置
  [SW5-HW-Vlanif1]ping 10.1.8.8 //测试与 PC 的数据通信，在防火墙开启的情况下可能实现单向通信，需关闭系统防火墙
    ping 10.1.8.8: 56 data bytes, press CTRL_C to break

```
Reply from 10.1.8.8: bytes=56 Sequence=1 ttl=64 time=1 ms
Reply from 10.1.8.8: bytes=56 Sequence=2 ttl=64 time=1 ms
Reply from 10.1.8.8: bytes=56 Sequence=3 ttl=64 time=1 ms
Reply from 10.1.8.8: bytes=56 Sequence=4 ttl=64 time=1 ms
Reply from 10.1.8.8: bytes=56 Sequence=5 ttl=64 time=1 ms

--- 10.1.8.8 ping statistics ---
  5 packet(s) transmitted
  5 packet(s) received
  0.00% packet loss
  round-trip min/avg/max = 1/1/1 ms
```

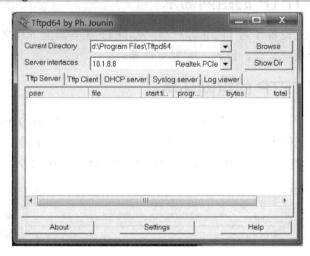

图 5-4 运行 TFTP 软件示意图

③ 配置华为设备的 TFTP 客户端功能。

```
<SW5-HW>dir    //基本的文件操作命令,该命令可列出所有文件和文件夹
Directory of flash:/

  Idx  Attr   Size(Byte)   Date         Time       FileName
    0  -rw-   25,107,548   Feb 08 2017  14:40:48   s5700-p-li-v200r009c00spc500.cc    //该文件为设
备的 VRP 系统文件
    1  drw-            -   Jul 25 2018  21:19:04   logfile
    2  drw-            -   Feb 08 2017  14:46:36   $_install_mod
    3  drw-            -   Feb 08 2017  14:46:46   user
    4  -rw-        1,740   Dec 17 2018  17:38:48   private-data.txt
    5  -rw-          836   Dec 17 2018  17:38:58   rr.dat
    6  -rw-          836   Dec 17 2018  17:38:58   rr.bak
    7  drw-            -   Feb 08 2017  14:47:38   localuser
    8  drw-            -   Feb 08 2017  14:47:38   dhcp
    9  -rw-            4   Jul 14 2017  19:30:16   snmpnotilog.txt
   10  -rw-          810   Dec 13 2018  16:06:46   vrpcfg.zip
   11  drw-            -   Jul 26 2018  00:00:02   resetinfo
   12  -rw-          122   Jul 11 2017  22:51:28   vrpcfg2.zip    //该文件为采用用户模式下
```

save 命令保存的配置文件

```
<SW5-HW>tftp 10.1.8.8 put vrpcfg2.zip     //把文件 vrpcfg2.zip 上传到 TFTP 服务器
Info: Transfer file in binary mode.
Uploading the file to the remote TFTP server. Please wait...
100%
TFTP: Uploading the file successfully.    //上传完成
122 byte(s) sent in 1 second(s).

<SW5-HW>delete flash:/vrpcfg2.zip    //删除文件
Delete flash:/vrpcfg2.zip?[Y/N]:y    //确定删除文件
Info: Deleting file flash:/vrpcfg2.zip...succeeded
<SW5-HW>tftp 10.1.8.8 get vrpcfg2.zip     //从服务器上下载文件
Info: Transfer file in binary mode.
Downloading the file from the remote TFTP server. Please wait...
100%
TFTP: Downloading the file successfully.   //文件下载成功
122 byte(s) received in 1 second(s).
```

由于 VRP 系统镜像（文件）较大，通过 TFTP 操控文件会耗时较长。请通过以下命令复制文件，如图 5-5 所示通过 TFTP 复制 VRP 系统镜像文件到 TFTP 服务器。

```
<SW5-HW>tftp 10.1.8.8 put s5700-p-li-v200r009c00spc500.cc    //把 VRP 系统文件上传到 TFTP 服务器的实施命令
```

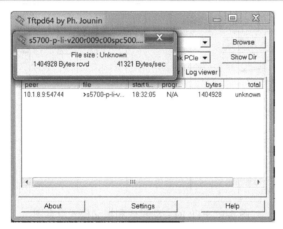

图 5-5　通过 TFTP 复制 VRP 系统镜像文件到 TFTP 服务器

其他升级 VRP 软件需要注意的事项：首先需要备份原有镜像文件，其次确保新的镜像文件与当前设备的硬件环境吻合（重要参数包括内存是否合适、Flash 空间是否满足等条件）。

至此，本案例实施完毕。

案例 6　配置 FTP

FTP（文件传送协议）简书

FTP（File Transfer Protocol，文件传送协议）是 TCP/IP 协议族中的一种应用层协议。FTP 的主要功能是向用户提供本地和远程主机之间的文件传输服务。在进行版本升级、日志下载和配置保存等业务操作时，会广泛地使用 FTP。FTP 采用两个 TCP 连接：控制连接和数据连接。其中，控制连接用于连接控制端口，传输控制命令；数据连接用于连接数据端口，传输数据。在控制连接建立后，数据连接通过控制端口的命令建立起连接，进行数据传输。FTP 数据连接的建立有两种模式：主动模式和被动模式，两者的区别在于数据连接是由服务器发起的还是由客户端发起的。图 6-1 所示为 FTP 的主动模式和被动模式示意图，读者可以参见图 6-1 对比两种模式的区别。

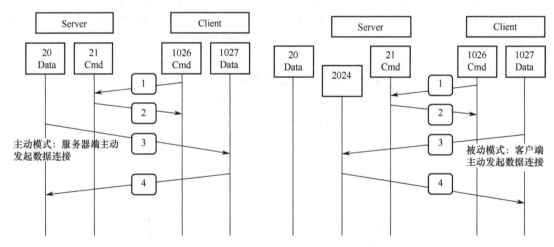

图 6-1　FTP 的主动模式和被动模式示意图

6.1　配置 FTP 案例拓扑

在本案例中，作为 FTP 服务器和客户端的设备采用华为的路由器和交换机，并且采用两台直连设备。在现网中，读者可以使用一些 FTP 软件（如 FileZilla），当然网络并不一定是直连网络，但要保证 FTP 服务器和 FTP 客户端可以通信。图 6-2 所示为在华为设备上配置 FTP 示意图。在图 6-2 中，AR1（地址为 192.168.0.1/30）作为 FTP 的服务器端，交换机 LSW1（VLAN1，地址为 192.168.0.2/30）作为 FTP 协议的客户端。

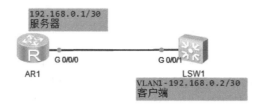

图 6-2 在华为设备上配置 FTP 示意图

6.2 使用 FTP 的组网需求

设备需要进行文件操作,将路由器作为 FTP 服务器,通过终端(交换机)将系统软件从路由器下载到本设备。该功能被普遍地用于 VRP 的备份、升级等操作中。

6.3 配置 FTP 步骤详解

配置 FTP 总体上可以按照以下 4 个步骤完成:
① 保证网络通信(本例采用最基本的直连方式通信)。
② 开启 FTP 服务。
③ 在 AAA 模式下设置正确的用户名、密码,以及该用户的权限、服务类型和共享的 TFP 目录。
④ 使用正确的测试方式。

6.3.1 配置路由器基本网络和 FTP 服务

① 首先须保证设备之间的网络可以通信。

```
<R1>sys    //进入系统视图
Enter system view, return user view with Ctrl+Z.
[R1]sysname AR-01    //修改设备名称为 AR-01
[AR-01]int g0/0/0    //进入 GigabitEthernet0/0/0 接口视图
[AR-01-GigabitEthernet0/0/0]ip address 192.168.0.2 30    //为该接口配置 IP 地址 192.168.0.2,掩码为 30(255.255.255.252)
IP on the interface GigabitEthernet0/0/0 has entered the UP state.
[AR-01-GigabitEthernet0/0/0]quit    //退出接口视图,返回系统视图
[AR-01]ping 192.168.0.2    //在本端测试是否可以 ping 通自己的地址,该步骤是为了验证本设备地址无误
  ping 192.168.0.2: 56    data bytes, press CTRL_C to break
  Reply from 192.168.0.2: bytes=56 Sequence=1 ttl=255 time=60 ms Reply from 192.168.0.2: bytes=56 Sequence=2 ttl=255 time=1 ms Reply from 192.168.0.2: bytes=56 Sequence=3 ttl=255 time=1 ms Reply from 192.168.0.2: bytes=56 Sequence=4 ttl=255 time=1 ms

  --- 192.168.0.2 ping statistics ---
  4 packet(s) transmitted
  4 packet(s) received
  0.00% packet loss
  round-trip min/avg/max = 1/15/60 ms
```

```
[AR-01]ping 192.168.0.1    //测试是否可以 ping 通对端地址（请先在交换机上配置地址）
ping 192.168.0.1: 56   data bytes, press CTRL_C to break
Reply from 192.168.0.1: bytes=56 Sequence=1 ttl=255 time=120 ms
Reply from 192.168.0.1: bytes=56 Sequence=2 ttl=255 time=10 ms
Reply from 192.168.0.1: bytes=56 Sequence=3 ttl=255 time=40 ms
Reply from 192.168.0.1: bytes=56 Sequence=4 ttl=255 time=20 ms
Reply from 192.168.0.1: bytes=56 Sequence=5 ttl=255 time=20 ms
--- 192.168.0.1 ping statistics ---
5 packet(s) transmitted
5 packet(s) received
0.00% packet loss
round-trip min/avg/max = 10/42/120 ms
```

② FTP 实施。

```
[AR-01]ftp server enable    //在路由器上开启 FTP 服务
Info: Succeeded in starting the FTP server
[AR-01]aaa    //进入 AAA 视图
[AR-01-aaa]local-user ender password cipher qytang001    //增加一个用户 ender，并设置密码为加密
模式 qytang001
Info: Add a new user.
[AR-01-aaa]local-user ender service-type ftp    //设置用户 ender 的服务类型为 ftp
[AR-01-aaa]local-user ender privilege level 15    //设置用户 ender 的权限为最高级别 15
[AR-01-aaa]local-user ender ftp-directory flash:    //设置用户 ender 的可操作目录为 flash，即所有目录
[AR-01-aaa]q    //退出 AAA 视图，返回系统视图
[AR-01]q    //退出系统视图，返回用户视图
<AR-01>save    //保存设置
The current configuration will be written to the device. Are you sure to continue? (y/n)[n]:y
It will take several minutes to save configuration file, please wait........
Configuration file had been saved successfully
Note: The configuration file will take effect after being activated
```

6.3.2 配置 FTP 客户端

本例中采用华为网络设备作为 FTP 的客户端，在现实工作中多采用一些 FTP 软件，比如 Filezilla 等。

```
[huawei]sysname LSW-01    //配置交换机名称为 LSW-01
[LSW-01]interface Vlanif1    //进入 Vlanif1 接口视图
[LSW-01-Vlanif1]ip address 192.168.0.1 30    //为该接口配置的地址为 192.168.0.1，掩码为 30
[LSW-01-Vlanif1]
Sep 21 2018 00:54:12-08:00 LSW-01 %%01IFNET/4/LINK_STATE(l)[1]:The line protocol
IP on the interface Vlanif1 has entered the Up State
```

测试 FTP 连接情况并复制文件。

```
<LSW-01>ftp 192.168.0.2    //在用户视图下通过 FTP 登录到 192.168.0.2 路由器
Trying 192.168.0.2 ... Press CTRL+K to abort Connected to 192.168.0.2.
220 FTP service ready. User(192.168.0.2:(none)):ender    //输入用户名 ender
331 Password required for ender. Enter password:    //输入对应的密码，注意密码并不显示
```

230 User logged in.

登录 FTP 服务器（AR1）之后显示如下：

[ftp]dir   //查看目录下文件及文件夹
200 Port command okay.
150 Opening ASCII mode data connection for *.
drwxrwxrwx   1 noone   nogroup   0 Sep 20 16:23 dhcp
-rwxrwxrwx   1 noone   nogroup   121802 May 26   2014 portalpage.zip
-rwxrwxrwx   1 noone   nogroup   2263 Sep 20 16:23 statemach.efs
-rwxrwxrwx   1 noone   nogroup   828482 May 26   2014 sslvpn.zip
-rwxrwxrwx   1 noone   nogroup   249 Sep 20 16:26 private-data.txt
drwxrwxrwx   1 noone   nogroup   0 Sep 20 16:28 .
-rwxrwxrwx   1 noone   nogroup   648 Sep 20 16:28 vrpcfg.zip
226 Transfer complete.

[ftp]get portalpage.zip   //通过 get 命令下载 portalpage.zip 文件
96%
99%
226 Transfer complete.
FTP: 121802 byte(s) received in 0.640 second(s) 190.31Kbyte(s)/sec.   //完成文件复制

至此，本案例实施完毕。

案例 7　网络设备文件系统管理

网络设备文件系统管理简书

在本案例中，我们将熟悉操控多种网络操作系统的文件管理命令，这些命令大部分时候类似于微软 Windows 系统中的查看、保存、复制等命令。由于大部分网络操作系统属于类 Linux 系统，所以很多命令都类似，华为与华三设备的这部分命令内容完全相同，所以本节中仅介绍华为的实施案例，读者可以参考华为的命令在华三设备上实施。华为的网络操作系统为通用路由平台（VRP）。VRP 的文件系统是 VRP 正常运行的基础，它负责管理存放于存储器上的系统文件（如 VRP 的 image）和配置文件等。现代流行的大多数网络操作系统是基于类 Linux 的系统，VRP 也不例外，我们可以在 VRP 的文件系统中对存储器中的文件、目录进行查看、创建、删除、修改等操作。掌握 VRP 文件系统的基本操作，可以帮助网络工程师对配置文件和 VRP 系统文件进行高效快速的管理。思科设备的 IOS（互联网操作系统）文件操作也非常类似。

7.1　文件系统管理案例拓扑

文件系统操作拓扑如图 7-1 所示，图中采用两台直连设备来完成此实验。两台直连设备的 IP 地址分别为 192.168.0.1/30 和 192.168.0.2/30。注意在 SW1 上使用 Vlanif1 接口，而非在 G0/0/1 接口上直接配置地址。

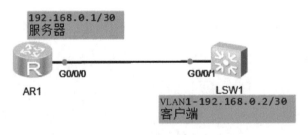

图 7-1　文件系统操作拓扑

7.2　文件管理需求

文件系统管理是指对存储器中文件、目录的管理，包括创建、删除、修改文件和目录，以及显示文件的内容。在本案例中，将对设备上的文件进行一系列的操作，如查看、创建、删除、修改文件或目录，具体需求请见 7.3 节的内容。

7.3 文件管理系统配置步骤详解

7.3.1 熟悉 VRP 系统的基本查看命令

```
<R1>dir    //显示当前目录下的文件信息,因为文件在 flash 根目录下,所以列出所有文件和文件夹
(d 代表文件夹)
Directory of flash:/

  Idx   Attr   Size(Byte)   Date          Time(LMT)    FileName
    0   drw-            -   Sep 22 2018   02:00:05     dhcp          //d 代表文件夹,r 代表可读权限,
w 代表可写权限
    1   -rw-      121,802   May 26 2014   09:20:58     portalpage.zip
    2   -rw-          555   Sep 22 2018   07:42:17     test.zip
    3   -rw-        2,263   Sep 22 2018   01:59:38     statemach.efs
    4   -rw-      828,482   May 26 2014   09:20:58     sslvpn.zip
    5   -rw-          249   Sep 22 2018   07:42:17     private-data.txt

1,090,732 KB total (784,456 KB free)
<R1>pwd    //查看当前所处的位置,当前位置为根目录
flash:
<R1>cd hcp    //一个错误的案例,由于输入了错误的文件夹,系统报错,不存在该文件或路径
Error: Such file or path doesn't exist    //该文件或路径不存在
<R1>cd dhcp    //进入 dhcp 文件夹
<R1>pwd    //查看光标当前所处位置
flash:/dhcp    //当前位于 flash 目录下的 dhcp 文件夹下
<R1>dir    //查看当前文件夹下的文件,可以看到 dhcp-duid.txt 文件
Directory of flash:/dhcp/

  Idx   Attr   Size(Byte)   Date          Time(LMT)    FileName
    0   -rw-           98   Sep 22 2018   02:00:05     dhcp-duid.txt

1,090,732 KB total (784,456 KB free)
<R1>more dhcp-duid.txt    //查看该文件的细节内容
*R1 DHCP DUID*
*time* 2018-09-22 10:00:05
*version* 1
#DUID_LL: 0003000100E0FCB90226
*end*
```

常用文件查看命令列表如表 7-1 所示。

表 7-1 常用文件查看命令列表

功能	命令	功能	命令
查看当前目录	pwd	查看文本文件的具体内容	more
显示当前目录下的文件信息	dir	修改用户当前界面的工作或进入新目录	cd
删除目录	rmdir	创建新目录	mkdir

7.3.2 配置 VRP 系统的目录

请按照下列操作熟悉系统文件操作：

```
<R1>dir        //查看 flash 根目录下的文件和文件夹
Directory of flash:/

 Idx  Attr    Size(Byte)  Date          Time(LMT)   FileName
  0   drw-         -      Sep 22 2018   09:55:49    dhcp
  1   -rw-    121,802     May 26 2014   09:20:58    portalpage.zip
  2   -rw-        555     Sep 22 2018   07:42:17    test.zip
  3   -rw-      2,263     Sep 22 2018   01:59:38    statemach.efs
  4   -rw-    828,482     May 26 2014   09:20:58    sslvpn.zip
  5   -rw-        249     Sep 22 2018   07:42:17    private-data.txt

<R1>cd dhcp    //进入 dhcp 文件夹
<R1>pwd        //查看当前所处位置，为 dhcp 目录
flash:/dhcp
<R1>mkdir test //在当前目录下创建 test 文件夹
Info: Create directory flash:/dhcp/test......Done
<R1>dir
Directory of flash:/dhcp/

 Idx  Attr    Size(Byte)  Date          Time(LMT)   FileName
  0   -rw-         98     Sep 22 2018   02:00:05    dhcp-duid.txt
  1   drw-          -     Sep 22 2018   09:55:20    test      //成功创建 test 文件夹

1,090,732 KB total (784,452 KB free)
<R1>rmdir test    //移除文件夹 test
Remove directory flash:/dhcp/test? (y/n)[n]:y   //此处会提示是否删除文件夹 test，请键入 y，确认删除。默认为否
%Removing directory flash:/dhcp/test...Done!
<R1>cd ..         //回到上级目录
<R1>pwd           //验证当前目录为 flash 根目录
flash:
```

7.3.3 配置 VRP 系统的文件

常用的文件操作命令包括复制、移动、重命名和删除文件等。

```
<R1>save ender.zip
Are you sure to save the configuration to ender.zip? (y/n)[n]:y   //保存当前配置为 ender.zip 的文件，键入 y 以确认
It will take several minutes to save configuration file, please wait........
Configuration file had been saved successfully
Note: The configuration file will take effect after being activated
<R1>copy ender.zip dhcp/
Copy flash:/ender.zip to flash:/dhcp/ender.zip? (y/n)[n]:y   //复制 ender/zip 文件到 dhcp 文件夹中
100%   complete
```

Info: Copied file flash:/ender.zip to flash:/dhcp/ender.zip...Done
<R1>cd dhcp/   //进入 dhcp 文件夹
<R1>dir   //通过 dir 命令看到 ender.zip 已经存在于 dhcp 目录下
Directory of flash:/dhcp/

```
  Idx   Attr   Size(Byte)   Date           Time(LMT)      FileName
   0    -rw-          98    Sep 22 2018    02:00:05       dhcp-duid.txt
   1    -rw-         557    Sep 22 2018    10:10:58       ender.zip
```

1,090,732 KB total (784,444 KB free)
<R1>delete ender.zip
Delete flash:/dhcp/ender.zip? (y/n)[n]:y   //删除 ender.zip，请键入 y 以确认删除
Info: Deleting file flash:/dhcp/ender.zip...succeed.
<R1>dir   //通过 dir 查看，得知文件已经被删除
Directory of flash:/dhcp/

```
  Idx   Attr   Size(Byte)   Date           Time(LMT)      FileName
   0    -rw-          98    Sep 22 2018    02:00:05       dhcp-duid.txt
```

1,090,732 KB total (784,444 KB free)
<R1>cd ..   //回到上级目录，即 flash 根目录
<R1>save qytang.cfg   //为了方便实验，保存当前运行的配置为文件名为 qytang.cfg 的文件
  Are you sure to save the configuration to qytang.cfg? (y/n)[n]:y   //键入 y 以确认
  It will take several minutes to save configuration file, please wait.......
  Configuration file had been saved successfully
  Note: The configuration file will take effect after being activated
<R1>rename qytang.cfg qyt.cfg   //通过 rename 命令把名为 qytang.zip 的文件修改为名为 qyt.cfg 的文件
Rename flash:/qytang.cfg to flash:/qyt.cfg? (y/n)[n]:y
Info: Rename file flash:/qytang.cfg to flash:/qyt.cfg ......Done
<R1>dir
Directory of flash:/

```
  Idx   Attr   Size(Byte)   Date           Time(LMT)      FileName
   0    drw-          -     Sep 22 2018    10:11:47       dhcp
   1    -rw-    121,802     May 26 2014    09:20:58       portalpage.zip
   2    -rw-        850     Sep 22 2018    10:12:56       qyt.cfg    //此时仅存在名为 qyt.cfg 的文件
   3    -rw-        555     Sep 22 2018    07:42:17       test.zip
   4    -rw-        557     Sep 22 2018    10:10:32       ender.zip
   5    -rw-      2,263     Sep 22 2018    01:59:38       statemach.efs
   6    -rw-    828,482     May 26 2014    09:20:58       sslvpn.zip
   7    -rw-        249     Sep 22 2018    07:42:17       private-data.txt
```

<R1>delete qyt.cfg
Delete flash:/qyt.cfg? (y/n)[n]:y   //使用 delete 命令删除文件 qyt.cfg。与 Windows 系统相同，该命令不会彻底删除文件，而是把文件移到回收站
Info: Deleting file flash:/qyt.cfg...succeed.
<R1>dir /all   //通过参数 all 可以查看所有文件，包括已删除的文件（以[]标识）

```
            Directory of flash:/

  Idx   Attr    Size(Byte)   Date         Time(LMT)   FileName
    0   drw-         -       Sep 22 2018  10:11:47    dhcp
    1   -rw-    121,802      May 26 2014  09:20:58    portalpage.zip
    2   -rw-        555      Sep 22 2018  07:42:17    test.zip
    3   -rw-        557      Sep 22 2018  10:10:32    ender.zip
    4   -rw-      2,263      Sep 22 2018  01:59:38    statemach.efs
    5   -rw-    828,482      May 26 2014  09:20:58    sslvpn.zip
    6   -rw-        249      Sep 22 2018  07:42:17    private-data.txt
    7   -rw-        850      Sep 22 2018  10:33:57    [qyt.cfg]

   <R1>reset recycle-bin qyt.cfg    //删除回收站中的 qyt.cfg 文件,如果不加文件名则删除回收站中的
所有文件
   Squeeze flash:/qyt.cfg? (y/n)[n]:y
   Clear file from flash will take a long time if needed...Done.
   %Cleared file flash:/qyt.cfg.
   <R1>move test.zip dhcp    //把文件 test.zip 移动到 dhcp 文件夹下,该命令类似于 Windows 系统中的剪
切命令
   Move flash:/test.zip to flash:/dhcp/test.zip? (y/n)[n]:y
```

常见文件操作命令汇总表如表 7-2 所示。

表 7-2 常见文件操作命令汇总表

功能	命令	功能	命令
复制文件	copy	删除/永久删除文件	Delete/unreserved
移动文件	move	恢复删除的文件	undelete
重命名文件	rename	彻底删除回收站中的文件	reset recycle-bin

7.3.4 管理 VRP 系统的配置文件

VRP 设备中的配置文件分为以下两种类型。
① 正在运行的配置文件。
② 保存的配置文件。

当前配置文件存储在设备的 RAM 中（类似于个人计算机的内存）。网络工程师的任务就是操控和优化网络系统 VRP 的配置，配置完成后使用 save 命令将当前配置保存到存储设备（Flash）中，形成保存的配置文件，这样 VRP 在下次启动之后就可以调用该配置文件来维护网络（就像我们修改完 Word 文件之后要保存到硬盘中）。保存的配置文件以 ".cfg" 或 ".zip" 作为扩展名（读者可以在 eNSP 的保存目录下看到类似的文件），存放在存储设备的根目录下。

初学者经常犯的错误就是配置完毕设备后没有保存，导致设备重启后出现空配置，这将严重影响后续工作。

在设备启动时，会从默认的存储路径加载保存的配置文件（saved-configuration file），并调用到 RAM 中。如果默认存储路径中没有保存的配置文件，则设备会使用默认参数进行初

案例 7　网络设备文件系统管理

始化配置。

```
<R1>save    //在用户视图下键入保存命令
  The current configuration will be written to the device.
  Are you sure to continue? (y/n)[n]:y    //在用户模式下的配置保存命令，键入 y 以确认
  It will take several minutes to save configuration file, please wait.......
  Configuration file had been saved successfully
  Note: The configuration file will take effect after being activated
<R1>dir
Directory of flash:/

  Idx   Attr    Size(Byte)   Date          Time(LMT)   FileName
    0   drw-           -     Sep 22 2018   10:53:59    dhcp
    1   -rw-     121,802     May 26 2014   09:20:58    portalpage.zip
    2   -rw-         555     Sep 22 2018   10:55:15    test.zip
    3   -rw-       2,263     Sep 22 2018   01:59:38    statemach.efs
    4   -rw-     828,482     May 26 2014   09:20:58    sslvpn.zip
    5   -rw-         249     Sep 22 2018   07:42:17    private-data.txt
    6   -rw-         559     Sep 22 2018   10:59:59    vrpcfg.zip    //默认将配置保存为文件名为
```
vrp.cfg 的文件

```
1,090,732 KB total (784,432 KB free)

<R1>system-view
Enter system view, return user view with Ctrl+Z.
[R1]sysname ENDER    //修改主机名为 ENDER

[ENDER]display current-configuration    //查看当前运行在内存中且还没保存的配置
[V200R003C00]
#
 sysname ENDER    //注意此时的主机名为 ENDER
#
 snmp-agent local-engineid 800007DB03000000000000
 snmp-agent
#
 clock timezone China-Standard-Time minus 08:00:00
#
portal local-server load portalpage.zip
#
 drop illegal-mac alarm
#
 set cpu-usage threshold 80 restore 75
#
aaa
 authentication-scheme default
 authorization-scheme default
 accounting-scheme default
 domain default
```

```
    domain default_admin
     local-user admin password cipher %$%$K8m.Nt84DZ}e#<0`8bmE3Uw}%$%$
     local-user admin service-type http
    #
```

[ENDER]display saved-configuration　　//此命令可查看已经保存的配置，请注意主机名为之前保存的 R1

```
[V200R003C00]
#
 sysname R1   //保存的配置文件中的主机名为 R1，如果不保存已经修改的内容，那么系统下次将
```
启用 R1 的主机名
```
#
 snmp-agent local-engineid 800007DB03000000000000
 snmp-agent
#
 clock timezone China-Standard-Time minus 08:00:00
#
 portal local-server load portalpage.zip
#
 drop illegal-mac alarm
#
 set cpu-usage threshold 80 restore 75
#
aaa
   authentication-scheme default
   authorization-scheme default
   accounting-scheme default
   domain default
   domain default_admin
   local-user admin password cipher %$%$K8m.Nt84DZ}e#<0`8bmE3Uw}%$%$
   local-user admin service-type http
#
```
[ENDER]interface g0/0/0
[ENDER-GigabitEthernet0/0/0]display this　　//华为设备上一个非常容易查看当前位置的命令是 display this，即查看当前位置的配置，此处是查看接口的配置。当然读者也不要一叶障目，相对于更先进的网络操作系统，VRP 系统的很多设置稍显古板，而且对实施顺序要求严格。当然其在一定程度上具有错误较少的优势，但在实施上确实非常麻烦
```
[V200R003C00]
#
interface GigabitEthernet0/0/0
 ip address 192.168.0.1 255.255.255.252
```
[ENDER]display current-configuration configuration user-interface
//VRP 也可以快速查询某些协议、接口、AAA、过滤等配置，此处为查看用户接口命令，显示结果如下
```
[V200R003C00]
#
user-interface con 0
```

```
    authentication-mode password
user-interface vty 0 4
user-interface vty 16 20
[ENDER]display current-configuration | include aaa    //华为VRP系统的管道符命令可以快速定位特
```
定配置，比如此处为"包含 aaa"的配置内容
```
aaa
[ENDER]display current-configuration | begin aaa    //华为VRP系统的管道符命令可以快速定位特定
```
配置，比如此处为"以 aaa 开始"的配置内容
```
aaa
    authentication-scheme default
    authorization-scheme default
    accounting-scheme default
    domain default
    domain default_admin
    local-user admin password cipher %$%$K8m.Nt84DZ}e#<0`8bmE3Uw}%$%$
    local-user admin service-type http
#
firewall zone Local
```

常用实施验证命令如表 7-3 所示。

表 7-3　常用实施验证命令

功能	命令
显示当前配置文件	display current-configuration
显示保存的配置文件	display saved-configuration
显示当前模式下的配置内容	display this
显示包含特定内容的配置	display current-configuration | include
显示以特定内容开始的配置	display current-configuration | begin
显示协议、用户接口、AAA 等配置	display current-configuration configuration

7.3.5　指定 VRP 系统启动文件和恢复出厂配置

系统启动配置文件，默认情况下会调用根目录下的启动文件，而当设备有备份配置文件时，可以指定其调用的配置文件名，这样可以灵活地实施项目。

```
[R1]sysname Ender-LAB
<Ender-LAB>save ender.cfg    //在用户模式下保存配置为文件名为 ender.cfg 的文件
Are you sure to save the configuration to flash:/ender.cfg?[Y/N]:
<Ender-LAB>dir    //查看结果表明该文件已经存在于 flash 中
Directory of flash:/

    Idx  Attr    Size(Byte)  Date          Time        FileName
    0    drw-    -           Mar 09 2016   17:01:50    src
    1    drw-    -           Mar 09 2016   17:01:54    pmdata
    2    drw-    -           Mar 09 2016   17:02:02    dhcp
    3    -rw-    28          Mar 09 2016   17:07:12    private-data.txt
    4    drw-    -           Mar 09 2016   17:22:18    mplstpoam
```

```
           5  -rw-              795  Mar 09 2016 17:22:46   ender.cfg
```

32,004 KB total (31,993 KB free)

<Ender-LAB>startup ?   //在这个位置我们可以看到的文件，其实在启动的时候可以指定多个参数，比如许可证、补丁、系统软件等参数

```
  license               Set license file
  paf                   Set paf file
  patch                 Set patch file
  saved-configuration   Saved-configuration file for system to startup
  system-software       Config system software for system to startup
```

<Ender-LAB>startup saved-configuacration ender.cfg   //指定启动配置文件名
Info: Succeeded in setting the configuration for booting system.
<Ender-LAB>save   //保存
The current configuration will be written to the device.
Are you sure to continue?[Y/N]y
Now saving the current configuration to the slot 17.
Mar 9 2016 17:23:58-08:00 Ender-LAB %%01CFM/4/SAVE(l)[1]:The user chose Y when deciding whether to save the configuration to the device.
Save the configuration successfully.
<Ender-LAB>reboot
Info: The system is now comparing the configuration, please wait.
Info: If want to reboot with saving diagnostic information, input 'N' and then execute 'reboot save diagnostic-information'.
System will reboot! Continue?[Y/N]:y   //重启之后我们将会看到设备从指定文件调用配置文件

<Ender-LAB>
<Ender-LAB>display startup
MainBoard:
```
  Configured startup system software:      NULL
  Startup system software:                 NULL
  Next startup system software:            NULL
  Startup saved-configuration file:        flash:/ender.cfg
  Next startup saved-configuration file:   flash:/ender.cfg    //下次启动时调用的配置文件
  Startup paf file:                        NULL
  Next startup paf file:                   NULL
  Startup license file:                    NULL
  Next startup license file:               NULL
  Startup patch package:                   NULL
```

恢复设备初始化的配置或者出厂配置（即完成从有配置到空配置），该行为的目的是从空白实施配置或从空白开始实验。读者可能会在实际操作时遇到大量的实验配置问题，所以这一部分的内容比较实用。

<QYT-LAB>reset saved-configuration   //清空保存的配置文件
Warning: The action will delete the saved configuration in the device.
The configuration will be erased to reconfigure. Continue? [Y/N]:Y   //该配置将会被擦除以重新配置，是否继续？请键入 Y
Warning: Now clearing the configuration in the device.

Error: The config file does not exist.
<QYT-LAB>
Mar　9 2016 17:06:40-08:00 QYT-LAB %%01CFM/4/RST_CFG(l)[0]:The user chose Y when deciding whether to reset the saved configuration.
<QYT-LAB>reboot　　//重新启动设备才能清空配置
Info: The system is now comparing the configuration, please wait.
Warning: All the configuration will be saved to the configuration file for the next startup:, Continue?[Y/N]:N　　//这里是非常容易犯错误的位置,"所有的配置将会保存到下次启动文件中,继续吗?"请一定键入 N, 否则无法得到想要的效果
Info: If want to reboot with saving diagnostic information, input 'N' and then execute 'reboot save diagnostic-information'.
System will reboot! Continue?[Y/N]:Y　　//"系统将会重启!继续?"请一定键入 Y 以确认
Mar　9 2016 17:06:55-08:00 QYT-LAB %%01CMD/4/REBOOT(l)[1]:The user chose Y when deciding whether to reboot the system. (Task=co0, Ip=**, User=**)

<ENDER>factory-configuration reset　　//该命令需要在真机上验证,在模拟器上不能恢复出厂的空配置
　Info: Successfully set factory config reset flag!

<ENDER>reboot
Info: The system is comparing the configuration, please wait.
Warning: All the configuration will be saved to the next startup configuration. Continue ? [y/n]:n
System will reboot! Continue ? [y/n]:y
Info: system is rebooting ,please wait...

7.3.6　熟悉思科网元的文件系统操作

熟悉 IOS 系统的基本查看命令,由于它也属于类 Linux 系统,所以读者会发现这些命令的实施与在 VRP 上的实施非常类似。请在思科设备上熟悉以下操作。

```
Router#dir    //显示当前目录下的文件信息,由于在系统下,即根目录下,所以列出所有文件和文件夹(d 代表文件夹)
Directory of system:/

    758  drwx       0         <no date>    cme
      2  -r--       0         <no date>    default-running-config
    766  dr-x       0         <no date>    fpm
      3  drwx       0         <no date>    its     //d 代表文件夹,r 代表可读权限,w 代表可写权限
    105  dr-x       0         <no date>    memory
      1  -rw-    1182         <no date>    running-config
    104  dr-x       0         <no date>    vfiles

No space information available
Router#pwd    //查看当前所处的位置,当前为根目录,思科模拟器会显示 system,在真实设备上会显示 flash
system:

Router#cd system:its    //cd 命令为进入文件夹命令,此命令完成进入 its 文件夹操作
Router#pwd    //查看当前所处位置
```

```
            system:/its/
            Router#cd ..      //回到上一级目录，".."代表上级目录
            Router#more system:/default-running-config    //查看系统目录下的文件名为 default-running-config 的
文件的详细内容
            !
            ! Last configuration change at 21:35:04 EET Tue Sep 25 2018
            version 15.3
            parser cache
            parser config partition
            parser command serializer
            parser maximum utilization 100
            parser maximum latency 40
            downward-compatible-config 15.3
            no service log backtrace
            no service config
            no service exec-callback
            no service nagle
            service slave-log
            no service slave-coredump
```

IOS 系统目录操作如下。

由于 EVE 模拟系统的问题，以下命令在模拟器上无效，但是在真机上有效。

```
            Router#mkdir disk0:
            Create directory filename []? TEST     //创建文件夹 TEST
            %Error Creating dir disk0:TEST (No such device)

            Router#rmdir      //删除文件夹 TEST（该命令不支持在模拟器上操作，在真机上可以成功操作）
            Remove directory filename []? TEST
            %Filesystem does not support rmdir operations
```

文件操作。

```
            Router#copy running-config system:test    //把当前运行的配置保存到名为 test 的文件中（模拟器不
支持，在真机上可以成功操作）
            Destination filename [test]?
            %Error opening system:test (No such file or directory)
```

这些 IOS 的命令与 VRP 的命令非常类似，读者可以灵活运用。

至此，本案例实施完毕。

第二篇 交换网络

在了解了第一篇的基础知识之后,本篇开始正式学习网络知识的交换部分内容,这部分内容是实际网络实施中应用最多的内容,作为一名网络工程师,VLAN、生成树、网关冗余备份协议、以太网链路聚合、设备堆叠、端口安全技术在维护园区以太网时都是必备的构建网络和维护网络的基本知识。

本篇内容包含案例8～27。

案例 8 交换机的 MAC 地址表

MAC 地址表简书

MAC 地址表就像一本通信录，记录了所有人的姓名和地址，通过这本通信录，你可以找到你想联系的人。交换机通过学习数据帧的源 MAC 地址构建这本通信录（MAC 地址表），在这本通信录中，不仅要记录源 MAC 地址，还需要同时记录 MAC 地址来源的端口，并将 MAC 地址和来源端口绑定。这样我们既知道了这个人的姓名（MAC 地址）也知道他的家庭住址（端口）。有了这本通信录以后，就可以转发数据帧了，交换机收到数据帧后，根据目的 MAC 地址（收件人姓名）查询通信录，找到收件人及其家庭住址（端口），将数据帧从相应端口转发出去。通过这个流程读者一定会觉得交换机就是个快递员啊！没错，它就是快递员，只不过快递的是数据帧。但是它也有它的苦恼，那就是交换机的通信录（MAC 地址表）能记录的信息数量是有限的，那如果之前记录过的地址（MAC 地址）已经过时了，不存在了，那交换机还需要继续在通信录中记录这个无效的地址（MAC 地址）吗？答案是不会的，交换机会为每一条 MAC 地址都启动一个定时器，定时器的过期时间为 300 秒（可以更改），如果在 300 秒内没有收到这个 MAC 地址发送的数据帧，则将这个 MAC 地址从通信录（MAC 地址表）中删除，留出位置记录其他的 MAC 地址。另外，一个 MAC 地址在同一时刻只能从一个端口得到，而不能从多个端口得到，但在一个端口下是可以学习到多个不同的 MAC 地址的。

下面通过图来具体了解交换机的转发原理。

PCA 发送数据帧给 PCB［目的 MAC 地址=PCD MAC 地址（0260.8c01.4444），源 MAC 地址=PCA MAC 地址（0260.8c01.1111）］，交换机从 E0 端口收到数据帧后，根据数据帧的目的 MAC 地址查找交换机的 MAC 地址表。由于交换机的 MAC 地址表为空，在 MAC 地址表中找不到目的 MAC 地址，交换机便将这个数据帧从接收端口（E0）以外的其他端口全部转发出去，这种特性叫作泛洪，该过程如图 8-1 所示。

同时，交换机会学习数据帧的源 MAC 地址，并将该 MAC 地址和接收端口记录到 MAC 地址表中，这种特性叫作学习（学习源 MAC 地址），该过程如图 8-2 所示。

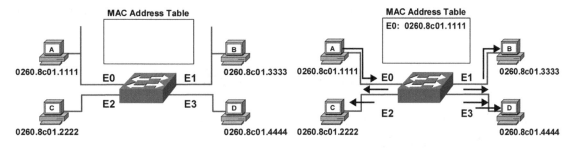

图 8-1 泛洪过程　　　　　　　　　　图 8-2 学习过程

由于这个数据帧被交换机泛洪出去，所以 PCB、PCC 和 PCD 都能收到该数据帧，但是 PCB 和 PCC 会丢弃这个数据帧，原因是它们接收数据帧后会对比数据帧的目的 MAC 地址和接收端口的 MAC 地址是否一致。如果一致则说明数据帧是发送到本设备的，便会接收；如果不一致则丢弃。PCD 收到数据帧后同样检查目的 MAC 地址是否和自己端口的 MAC 地址一致，若发现 MAC 地址一致，则 PCD 知道数据帧是发送给自己的，便拆封二层数据帧交给上层处理。同时向 PCA 发送一个回复数据帧［目的 MAC 地址=PCA MAC 地址（0260.8c01.1111），源 MAC 地址=PCD MAC 地址（0260.8c01.4444）］。

交换机从 E1 端口收到回复数据帧，根据数据帧的目的 MAC 地址查找交换机的 MAC 地址表，交换机在 MAC 地址表中查找到了目的 MAC 地址，并且这个 MAC 地址被绑定到了 E0 端口上，交换机便将这个数据帧直接从 E0 端口转发出去，而不再进行泛洪，通过已有 MAC 地址表转发数据帧过程如图 8-3 所示。

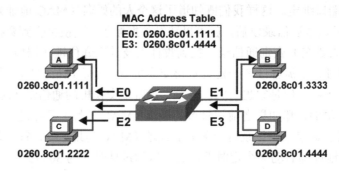

图 8-3 通过已有 MAC 地址表转发数据帧过程

8.1 查验 MAC 地址表网络拓扑

本章拓扑请在后续三个分解拓扑中查看。基本拓扑是一台交换机连接三台终端，三台终端的 IP 地址分别为 10.1.10.1/28、10.1.10.2/28 和 10.1.10.3/28，这是基本的网络设计。

8.2 学习 MAC 地址表的现实需求

在思科、华为、华三设备上观察 MAC 地址表，并熟悉 MAC 地址表各个表项中所包含的信息，以便日后在交换机上进行排错，也为后续学习端口安全技术做好知识储备。需要注意后面会学习 Trunk、Hybrid 等交换机的链路类型，在这些场景下，从一个端口可以学习到多个 MAC 地址。

8.3 验证 MAC 地址表的要点

① 配置终端 IP 地址并发送数据，从而触发 MAC 地址表的学习。
② 修改 MAC 地址表老化时间配置。

8.4 验证 MAC 地址表的实施步骤

8.4.1 在华三设备上学习 MAC 地址表

在华三设备上学习 MAC 地址表拓扑如图 8-4 所示，本节案例依照图 8-4 实施。

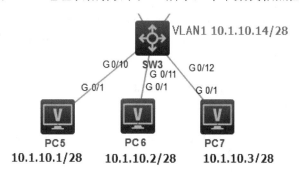

图 8-4　在华三设备上学习 MAC 地址表拓扑

图 8-5 所示为华三终端模拟器 IP 地址配置示意图，在华三模拟器（HCL）上配置 PC 地址的操作如该图所示。

图 8-5　华三终端模拟器 IP 地址配置示意图

- 右键单击 PC，选择"配置"，在"接口管理"处选择启用，即❶处；
- 选择 IPv4 配置中的"静态"，即❷处；
- 配置对应的 IPv4 地址、掩码地址和 IPv4 网关（本例中网关并不是必要的配置），即❸处；
- 单击"启用"，即❹处。

在华三交换机上实施和查看，如下所示。

```
[SW3]display mac-address      //初始情况下 MAC 地址表为空
MAC Address        VLAN ID       State          Port/Nickname         Aging

[SW3]interface Vlan-interface 1
[SW3-Vlan-interface1]ip address 10.1.10.14 28   //在交换机上配置 Vlanif 接口，默认情况下交换机上的所有接口属于 VLAN1，此时可以配置管理 VLAN1，并赋予其一个 IP 地址，即可与 VLAN1 下的终端通信

[SW3]ping 10.1.10.1    //在 SW3 上主动向终端发送报文，这个操作的目的是主动在网络中产生流量，之后 MAC 地址就可以自动进行学习了
ping 10.1.10.1 (10.1.10.1): 56 data bytes, press CTRL_C to break
56 bytes from 10.1.10.1: icmp_seq=0 ttl=255 time=5.000 ms
56 bytes from 10.1.10.1: icmp_seq=1 ttl=255 time=5.000 ms
56 bytes from 10.1.10.1: icmp_seq=2 ttl=255 time=2.000 ms
56 bytes from 10.1.10.1: icmp_seq=3 ttl=255 time=1.000 ms
56 bytes from 10.1.10.1: icmp_seq=4 ttl=255 time=1.000 ms
[SW3]ping 10.1.10.2
ping 10.1.10.2 (10.1.10.2): 56 data bytes, press CTRL_C to break
56 bytes from 10.1.10.2: icmp_seq=0 ttl=255 time=7.000 ms
56 bytes from 10.1.10.2: icmp_seq=1 ttl=255 time=4.000 ms
56 bytes from 10.1.10.2: icmp_seq=2 ttl=255 time=4.000 ms
56 bytes from 10.1.10.2: icmp_seq=3 ttl=255 time=3.000 ms
56 bytes from 10.1.10.2: icmp_seq=4 ttl=255 time=26.000 ms
[SW3]ping 10.1.10.3
ping 10.1.10.3 (10.1.10.3): 56 data bytes, press CTRL_C to break
56 bytes from 10.1.10.3: icmp_seq=0 ttl=255 time=6.000 ms
56 bytes from 10.1.10.3: icmp_seq=1 ttl=255 time=2.000 ms
56 bytes from 10.1.10.3: icmp_seq=2 ttl=255 time=2.000 ms
56 bytes from 10.1.10.3: icmp_seq=3 ttl=255 time=1.000 ms
56 bytes from 10.1.10.3: icmp_seq=4 ttl=255 time=3.000 ms

<SW3>display mac-address   //SW3 已经学习到了 3 台终端的 MAC 地址，它们都属于 VLAN1，状态为"学习到"，分别对应图 8-4 中的 G0/10、G0/11 和 G0/12，都在进行老化计时（即如果 300 秒内没有流量，该 MAC 地址表项会在 300 秒后消失，除非有流量再次触发它）
MAC Address        VLAN ID       State          Port/Nickname         Aging
8406-dddc-0506        1          Learned        G1/0/10               Y
8406-e464-0606        1          Learned        G1/0/11               Y
8407-0d36-0706        1          Learned        G1/0/12               Y

[SW3]mac-address timer aging 299    //MAC 地址表的老化时间默认为 300 秒，此处修改为 299 秒，通常不建议修改 MAC 地址表的老化时间
[SW3]display mac-address aging-time
MAC address aging time: 299s.
```

8.4.2 在思科设备上学习 MAC 地址表

图 8-6 为思科设备学习 MAC 地址表拓扑，本节案例依照图 8-6 所示实施。

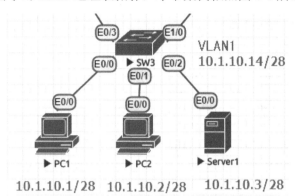

图 8-6 思科设备学习 MAC 地址表拓扑

在交换机上实施如下操作。

```
Switch(config)#hostname SW3
SW3(config)#interface vlan 1
SW3(config-if)#no sh    //在思科设备上三层接口默认是关闭状态，需要读者手动开启接口
SW3(config-if)#ip address 10.1.10.14 255.255.255.240    //思科设备上不能直接配置"/x"这种类似华为、
华三的掩码形式，只能配置十进制的掩码
SW3(config-if)#exit
SW3#show mac address-table    //默认情况下 MAC 地址表为空状态
          Mac Address Table
-------------------------------------------

Vlan    Mac Address     Type      Ports
----    -----------     --------  -----
```

请读者给拓扑中的 PC1、PC2 和 Server1 对应接口自行配置地址（不要忘记开启接口）。

```
SW3#ping 10.1.10.1    //在交换机上主动发送到达各个终端的流量以触发 MAC 地址表学习
Type escape sequence to abort.
Sending 5, 100-byte ICMP Echos to 10.1.10.1, timeout is 2 seconds:
.!!!!
Success rate is 80 percent (4/5), round-trip min/avg/max = 1/1/3 ms
SW3#ping 10.1.10.2
Type escape sequence to abort.
Sending 5, 100-byte ICMP Echos to 10.1.10.2, timeout is 2 seconds:
.!!!!
Success rate is 80 percent (4/5), round-trip min/avg/max = 1/1/2 ms
SW3#ping 10.1.10.3
Type escape sequence to abort.
Sending 5, 100-byte ICMP Echos to 10.1.10.3, timeout is 2 seconds:
```

```
.!!!!
Success rate is 80 percent (4/5), round-trip min/avg/max = 1/1/2 ms
SW3#show mac address-table    //验证思科设备上的 MAC 地址表，该表项包含了所属的 VLAN、终
端的 MAC 地址、类型，以及动态学习、从哪个端口学习到 MAC 地址
              Mac Address Table
-------------------------------------------

Vlan    Mac Address       Type         Ports
----    -----------       --------     -----
 1      aabb.cc00.8000    DYNAMIC      E0/0
 1      aabb.cc00.9000    DYNAMIC      E0/1
 1      aabb.cc00.b000    DYNAMIC      E0/2
Total Mac Addresses for this criterion: 3
SW3(config)#mac address-table aging-time ?
  <0-0>              Enter 0 to disable aging
  <10-1000000>       Aging time in seconds

SW3(config)#mac address-table aging-time 299    //修改 MAC 地址表的老化时间为 299 秒，默认为
300 秒
SW3(config)#do show mac address aging
Global Aging Time:    299
Vlan      Aging Time
----      ----------
```

在思科设备上学习 MAC 地址表案例实施完毕。

8.4.3　在华为设备上学习 MAC 地址表

在华为设备上学习 MAC 地址表拓扑如图 8-7 所示，本节案例依照图 8-7 实施。

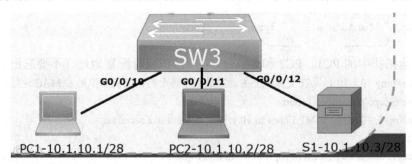

图 8-7　在华为设备上学习 MAC 地址表拓扑

配置 eNSP 终端的 IP 地址如图 8-8 所示。首先双击终端图标，然后按照图 8-8 所示配置 IP 地址、子网掩码和网关，最后不要忘记单击"应用"，华为的 eNSP 的服务器的配置与 PC 类似，请读者参考。eNSP 服务器的基本配置如图 8-9 所示。

案例 8 交换机的 MAC 地址表

图 8-8 配置 eNSP 终端的 IP 地址

图 8-9 eNSP 服务器的基本配置

在交换机上实施如下验证操作。

```
[huawei]sysname SW3
[SW3]interface Vlanif 1
[SW3-Vlanif1]ip address 10.1.10.14 28    //在华为设备上配置 VLAN 的地址，华为和华三设备一样可以直接配置掩码的长度
[SW3-Vlanif1]display mac-address    //默认情况下，交换机上的 MAC 地址表为空
[SW3-Vlanif1]
```

[SW3-Vlanif1]ping 10.1.10.1   //主动向终端发送数据以触发 MAC 地址学习

　ping 10.1.10.1: 56　　data bytes, press CTRL_C to break
　　Reply from 10.1.10.1: bytes=56 Sequence=1 ttl=128 time=70 ms
　　Reply from 10.1.10.1: bytes=56 Sequence=2 ttl=128 time=30 ms
　　Reply from 10.1.10.1: bytes=56 Sequence=3 ttl=128 time=10 ms
　　Reply from 10.1.10.1: bytes=56 Sequence=4 ttl=128 time=30 ms
　　Reply from 10.1.10.1: bytes=56 Sequence=5 ttl=128 time=50 ms

　--- 10.1.10.1 ping statistics ---
　　5 packet(s) transmitted
　　5 packet(s) received
　　0.00% packet loss
　　round-trip min/avg/max = 10/38/70 ms

[SW3-Vlanif1]ping 10.1.10.2
　ping 10.1.10.2: 56　　data bytes, press CTRL_C to break
　　Reply from 10.1.10.2: bytes=56 Sequence=1 ttl=128 time=30 ms
　　Reply from 10.1.10.2: bytes=56 Sequence=2 ttl=128 time=40 ms
　　Reply from 10.1.10.2: bytes=56 Sequence=3 ttl=128 time=30 ms
　　Reply from 10.1.10.2: bytes=56 Sequence=4 ttl=128 time=30 ms
　　Reply from 10.1.10.2: bytes=56 Sequence=5 ttl=128 time=50 ms

　--- 10.1.10.2 ping statistics ---
　　5 packet(s) transmitted
　　5 packet(s) received
　　0.00% packet loss
　　round-trip min/avg/max = 30/36/50 ms

[SW3-Vlanif1]ping 10.1.10.3
　ping 10.1.10.3: 56　　data bytes, press CTRL_C to break
　　Reply from 10.1.10.3: bytes=56 Sequence=1 ttl=255 time=20 ms
　　Reply from 10.1.10.3: bytes=56 Sequence=2 ttl=255 time=10 ms
　　Reply from 10.1.10.3: bytes=56 Sequence=3 ttl=255 time=1 ms
　　Reply from 10.1.10.3: bytes=56 Sequence=4 ttl=255 time=1 ms
　　Reply from 10.1.10.3: bytes=56 Sequence=5 ttl=255 time=30 ms

　--- 10.1.10.3 ping statistics ---
　　5 packet(s) transmitted
　　5 packet(s) received
　　0.00% packet loss
　　round-trip min/avg/max = 1/12/30 m
[SW3]display mac-address   //华为设备的 MAC 地址表显示的内容和其他厂商类似，分别为终端的 MAC 地址、所属 VLAN、对应端口、动态方式学习等内容
　MAC address table of slot 0:

MAC Address　　　VLAN/　　　PEVLAN CEVLAN Port　　　　　Type　　　LSP/LSR-ID

案例 8　交换机的 MAC 地址表

```
                    VSI/SI                                    MAC-Tunnel
------------------------------------------------------------------------
5489-98b1-7b52   1         -         -        G0/0/10        dynamic    0/-
5489-9841-1c54   1         -         -        G0/0/11        dynamic    0/-
5489-9883-7156   1         -         -        G0/0/12        dynamic    0/-
------------------------------------------------------------------------
Total matching items on slot 0 displayed = 3
[SW3]mac-address aging-time ?
   <0,10-1000000>   Aging-time seconds, 0 means that MAC aging function does not
                    work

[SW3]mac-address aging-time 299    //修改华为设备 MAC 地址表的老化时间，从默认的 300 秒
改为 299 秒
[SW3]display mac-address aging-time
   Aging time: 299 second
```

至此，在华为设备上学习 MAC 地址表案例实施完毕。

案例 9　VLAN 技术接入案例

VLAN 和交换机接入端口简书

VLAN 的工作原理实际上是对 MAC 地址表的划分，每一个 VLAN 都会对应一张本 VLAN 的 MAC 地址表，通过将物理端口划分到不同的 VLAN 中，实现广播域（MAC 地址表）的隔离。默认情况下交换机的所有端口都属于 VLAN1，那么在同一 VLAN 内，交换机上连接的所有主机也都属于 VLAN1，属于同一个广播域。这会导致广播域过大，所以在现网中常划分为多个 VLAN 从而使得广播域变小，使得网络更便于管理。VLAN 实施背景示意图如图 9-1 所示，图中划分了多个 VLAN，而不是让所有用户都处于默认的 VLAN1。

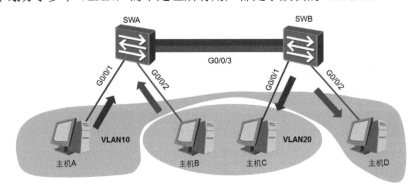

图 9-1　VLAN 实施背景示意图

VLAN 技术可以将一个物理局域网在逻辑上划分成多个广播域，也就是多个 VLAN。VLAN 技术部署在数据链路层，用于隔离二层流量。同一个 VLAN 内的主机共享同一个广播域，它们之间可以直接在二层互相通信。而 VLAN 间的主机属于不同的广播域，不能直接在二层互相通信。这样，广播报文就被限制在各个相应的 VLAN 内，同时也提高了网络安全性。

本案例中，原本属于同一广播域的主机被划分到了两个 VLAN 中，即 VLAN10 和 VLAN20。VLAN 内部的主机可以直接在二层互相通信，VLAN10 和 VLAN20 之间的主机无法直接实现二层通信。

VLAN 不是为了隔离用户，而是让用户在通信的同时可以被更好地管理。

一个 VLAN 即一个广播域，VLAN 内部的主机可以直接在二层互相通信，而 VLAN 间主机可以通过三层（路由技术）通信。

经典情况下，用户主机和交换机之间的链路为接入链路，交换机与交换机之间的链路为干道链路，这两种链路类型是通用类型。在不同的组网解决方案中可以实施不同的技术，最本质的内容为是否需要多个 VLAN ID 的流量经过端口转发。

华为、华三设备的 VLAN 端口类型有 4 种：Access 端口、Trunk 端口、Hybrid 端口（华为、华三以及一些厂商设备的特有端口），以及本书不涉及的 QinQ 模式（请关注笔者的 HCIP 及 HCIE 书籍）。

思科设备上的 VLAN 端口类型有 4 种：Access 端口、Trunk 端口、QinQ 端口和私有 VLAN 模式（Private VLAN）。私有 VLAN 模式端口类型在出入方向上只能允许唯一的 VLAN ID（帧）通过本端口，即独享 1 个 VLAN 的端口类型。Access 连接：端口发出去的报文不携带 VLAN TAG，一般用于与不能识别 VLAN TAG 的终端相连，或者在不需要区分不同 VLAN 流量时使用。携带 TAG 和不携带 TAG 的以太网帧结构如图 9-2 所示。

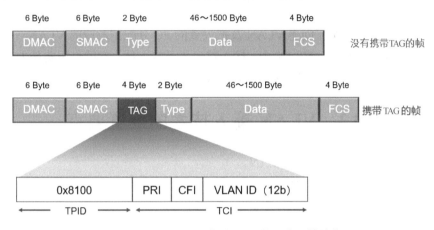

图 9-2　携带 TAG 和不携带 TAG 的以太网帧结构

Access 端口模式的工作原理如图 9-3 所示，在主机 A 访问主机 C PVID 的方向上，不携带任何标识的以太二型（Ethernet II）数据帧，在❶处，在进入交换机的 G0/0/1 端口时，会强制加上该端口的 VLAN ID（PVID）10（VLAN10 即代表该端口接入 VLAN10），那么，此时这些帧开始携带 VLAN 标识，即 TAG10。

在图 9-3 中的❷处，G0/0/3 也被接入 VLAN10，这意味着允许 VLAN10 的流量从该端口转发（即 VLAN List 允许通行），而帧的行为是去掉 TAG，把原始的以太网帧发送到主机 C。而该帧不能从 G0/0/2 转发出去，因为从该端口接入 VLAN2，这意味着允许通过的 VLAN 列表只有 VLAN2，而没有 VLAN10。

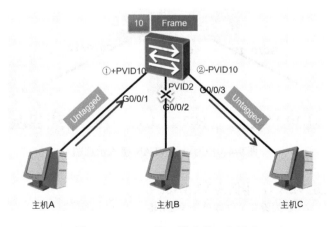

图 9-3　Access 端口模式的工作原理

9.1　实施端口接入 VLAN 的组网拓扑

在本案例中，我们在 3 个厂商的设备上实施基本的 VLAN 接入，所以请根据 9.4.1、9.4.2 和 9.4.3 节中各个拓扑来实现。

9.2　实施端口接入 VLAN 的组网需求

在本案例中，将介绍如何创建 VLAN 及把大量的 PC 分配到对应的 VLAN 中去。将 PC1、PC2 和 Server1 都划分到一个 VLAN 中，接入端口属于接入层，它的作用就是大量接入终端，同时，由于 3 个终端在同一子网，因此可以实现基本的通信。

9.3　VLAN 接入技术的配置要点

① 在交换设备上创建对应的 VLAN。
② 将对应端口的模式修改为接入模式。
③ 将端口划分到对应的 VLAN 中。
④ 使用终端进行数据测试。

9.4　VLAN 接入技术实施步骤详解

9.4.1　在华为设备上实施 VLAN 接入

图 9-4 为华为设备 VLAN 接入示意图，通过在图中的 SW3 上配置基本 VLAN 接入来介绍交换机如何增加端口 VLAN 标识（PVID），之后在同一交换机上，只能把数据帧从允许通过该 VLAN 的端口剥离掉该标识后，将原始帧从端口发送出去。

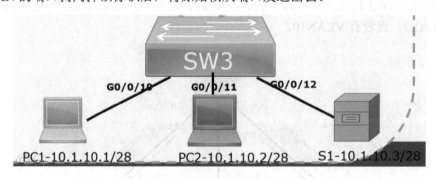

图 9-4　华为设备 VLAN 接入示意图

① 在设备上创建 VLAN，本例为 VLAN8。

```
[SW3]vlan 8    //在华为设备上创建单个 VLAN8
[SW3-vlan8]description QCNA    //华为设备不提供 VLAN 名，当然"描述"命令其实也是一个别名
```

[SW3]display port vlan active   //在华为设备上查看 VLAN 和链路类型（默认为 hybrid）及所属的 VLAN。读者可以看到，在 PVID（端口 VLAN 标识）一列默认端口都是 VLAN1，VLAN List 一列代表允许哪些 VLAN 的流量在去掉 TAG 之后从该端口发出去，默认也是 VLAN1

T=TAG U=UNTAG

Port	Link Type	PVID	VLAN List
G0/0/1	hybrid	1	U: 1
G0/0/2	hybrid	1	U: 1
G0/0/3	hybrid	1	U: 1
G0/0/4	hybrid	1	U: 1
G0/0/5	hybrid	1	U: 1
G0/0/6	hybrid	1	U: 1
G0/0/7	hybrid	1	U: 1
G0/0/8	hybrid	1	U: 1
G0/0/9	hybrid	1	U: 1
G0/0/10	hybrid	1	U: 1
G0/0/11	hybrid	1	U: 1
G0/0/12	hybrid	1	U: 1
G0/0/13	hybrid	1	U: 1
G0/0/14	hybrid	1	U: 1
G0/0/15	hybrid	1	U: 1
G0/0/16	hybrid	1	U: 1
G0/0/17	hybrid	1	U: 1
G0/0/18	hybrid	1	U: 1
G0/0/19	hybrid	1	U: 1
G0/0/20	hybrid	1	U: 1
G0/0/21	hybrid	1	U: 1
G0/0/22	hybrid	1	U: 1
G0/0/23	hybrid	1	U: 1
G0/0/24	hybrid	1	U: 1

[SW3]port-group group-member GigabitEthernet 0/0/10 to GigabitEthernet 0/0/12   //创建一个邻居端口组，其功能类似于思科的 rang 命令，当退出端口组之后，该端口组自动消失

[SW3-port-group]port link-type access

[SW3-GigabitEthernet0/0/10]port link-type access   //华为设备上默认的端口模式为 hybrid，把端口模式改为接入模式

[SW3-GigabitEthernet0/0/11]port link-type access

[SW3-GigabitEthernet0/0/12]port link-type access

[SW3-GigabitEthernet0/0/10]port default vlan 8   //端口所属的 VLAN 默认为 VLAN1，当前命令将其改为 VLAN8，该命令意味着流量从该端口进入时被增加了 PVID 8，同时仅允许 VLAN8 的流量在去掉 TAG 之后从该端口转发出去

[SW3-GigabitEthernet0/0/11]port default vlan 8

[SW3-GigabitEthernet0/0/12]port default vlan 8

[SW3-port-group]

[SW3-port-group]dis th

```
#
return
```
[SW3-port-group]display port vlan active   //验证端口的 VLAN 归属（类似于思科的 show vlan brief 命令），读者可以看到 G0/01/10、G0/01/11、G0/01/12 端口链路类型改为了 access 模式，默认的 PVID 修改为 8（命令 port default vlan 8 实现该功能），VLAN 列表中仅允许了 VLAN8 的流量通过（这是笔者经常在课程中讲的，access 模式是一种"独享模式"）

T=TAG U=UNTAG

Port	Link Type	PVID	VLAN List
G0/0/1	hybrid	1	U: 1
G0/0/2	hybrid	1	U: 1
G0/0/3	hybrid	1	U: 1
G0/0/4	hybrid	1	U: 1
G0/0/5	hybrid	1	U: 1
G0/0/6	hybrid	1	U: 1
G0/0/7	hybrid	1	U: 1
G0/0/8	hybrid	1	U: 1
G0/0/9	hybrid	1	U: 1
G0/0/10	access	8	U: 8
G0/0/11	access	8	U: 8
G0/0/12	access	8	U: 8
G0/0/13	hybrid	1	U: 1
G0/0/14	hybrid	1	U: 1
G0/0/15	hybrid	1	U: 1
G0/0/16	hybrid	1	U: 1
G0/0/17	hybrid	1	U: 1
G0/0/18	hybrid	1	U: 1
G0/0/19	hybrid	1	U: 1
G0/0/20	hybrid	1	U: 1
G0/0/21	hybrid	1	U: 1
G0/0/22	hybrid	1	U: 1
G0/0/23	hybrid	1	U: 1
G0/0/24	hybrid	1	U: 1

② 测试属于同一 VLAN 的流量，完成实验。

如果读者还保留之前案例 8 中 SW3 的 VLAN1 的地址，那么实施过程如下所示。

```
[SW3]display ip int brief
*down: administratively down
^down: standby
(l): loopback
(s): spoofing
The number of interface that is UP in Physical is 2
The number of interface that is DOWN in Physical is 1
The number of interface that is UP in Protocol is 2
The number of interface that is DOWN in Protocol is 1
```

Interface	IP Address/Mask	Physical	Protocol
MEth0/0/1	unassigned	down	down
NULL0	unassigned	up	up(s)
Vlanif1	10.1.10.14/28	up	up

读者可以思考一下，此时 VLAN1 的地址可以和 PC1 通信吗？
结果如下：

    [SW3]ping 10.1.10.1 //从 VLAN1 发送的帧只能发送到 VLAN1 中，而 PC1 属于 VLAN8，所以以太网帧不能从 G0/0/10 端口发送出去（VLAN List 起到了过滤作用）
      ping 10.1.10.1: 56 data bytes, press CTRL_C to break
      Request time out
      Request time out
      Request time out
      Request time out
      Request time out

      --- 10.1.10.1 ping statistics ---
      5 packet(s) transmitted
      0 packet(s) received
      100.00% packet loss

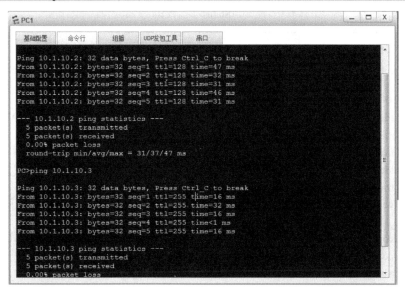

9.4.2 在思科设备上实施 VLAN 接入

思科设备 VLAN 接入示意图如图 9-5 所示。首先要创建 VLAN，通过在 SW3 上配置基本的 VLAN 接入来帮助读者理解交换机如何增加端口 VLAN 标识（PVID），之后在同一交换机上，只能把数据帧从允许该 VLAN 流量通过的端口剥离掉该标识后，以原始帧的形式从端口发送出去。

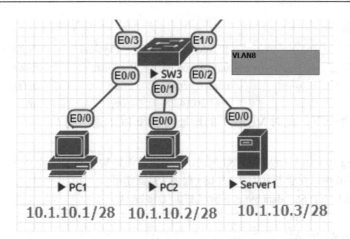

图 9-5 思科设备 VLAN 接入示意图

请在 SW3 上完成以下配置。

① 创建 VLAN，如果设备不存在 VLAN，不会转发对应 VLAN 的流量。

```
SW3(config)#vlan 8          //以自定义方式创建 VLAN8，而不使用默认的 VLAN1
SW3(config-vlan)#name QCNA  //将该 VLAN 命名为 QCNA
SW3#show vlan brief         //查看 VLAN 的简要信息

VLAN Name                             Status    Ports
---- -------------------------------- --------- -------------------------------
1    default                          active    Et0/0, Et0/1, Et0/2, Et0/3
                                                Et1/0, Et1/1, Et1/2, Et1/3
8    QCNA                             active    //VLAN8 已经创建成功，但当前该 VLAN 下没有
端口，VLAN 的作用是转发数据，没有端口的 VLAN 是没有意义的
1002 fddi-default                     act/unsup //思科设备上 VLAN1002 到 VLAN1005 是保留给
一个历史上曾经存在过的其他二层网络的，华为和华三设备上并没有保留这些 VLAN。
1003 token-ring-default               act/unsup
1004 fddinet-default                  act/unsup
1005 trnet-default                    act/unsup
SW3(config)#int range e0/0 – 2  //同时进入 e0/0 到 e0/2 端口，一并完成配置。中间不能有间隔的
                                端口
SW3(config-if-range)#switchport mode access   //将交换端口模式配置为 access 模式
SW3(config-if-range)#switchport access vlan 8 //端口被静态分配到 VLAN8 中
SW3#show vlan brief

VLAN Name                             Status    Ports
---- -------------------------------- --------- -------------------------------
1    default                          active    E0/3, E1/0, E1/1, E1/2
                                                E1/3
8    QCNA                             active    E0/0, E0/1, E0/2   //3 个端口已经被划分到 VLAN8
中，那么只要这些终端的 IP 地址属于同一子网就可以实现通信
1002 fddi-default                     act/unsup
1003 token-ring-default               act/unsup
1004 fddinet-default                  act/unsup
```

```
1005 trnet-default                    act/unsup
```
在终端上测试三层通信情况。
```
PC1#ping 10.1.10.2    //PC1 可以与 PC2 实现通信
Type escape sequence to abort.
Sending 5, 100-byte ICMP Echos to 10.1.10.2, timeout is 2 seconds:
.!!!!
Success rate is 80 percent (4/5), round-trip min/avg/max = 1/2/4 ms
PC1#ping 10.1.10.3    //PC1 可以与 PC3 通信

Type escape sequence to abort.
Sending 5, 100-byte ICMP Echos to 10.1.10.3, timeout is 2 seconds:
.!!!!
Success rate is 80 percent (4/5), round-trip min/avg/max = 2/9/29 ms
```
② 思科设备的基本 VLAN 创建和端口接入 VLAN 实施成功。

9.4.3 在华三设备上实施 VLAN 接入

图 9-6 所示为华三设备 VLAN 接入示意图,图中演示了网络接入交换机的接入用户功能,这是网络的一个部分,终端业务用户通过该功能接入企业网络。

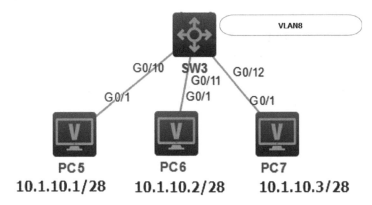

图 9-6 华三设备 VLAN 接入示意图

本案例实施请参照图 9-4 所示拓扑。

```
[SW3]display vlan brief    //查看华三设备上默认的 VLAN 归属情况,读者可以看到,默认情况下
                             所有厂商设备端口都属于 VLAN1,请注意"vlan1"不能删除
Brief information about all VLANs:
Supported Minimum VLAN ID: 1
Supported Maximum VLAN ID: 4094
Default VLAN ID: 1
VLAN ID    Name          Port
1          VLAN 0001     FG1/0/53    FG1/0/54    G1/0/1
                         G1/0/2      G1/0/3      G1/0/4    G1/0/5
                         G1/0/6      G1/0/7      G1/0/8    G1/0/9
                         G1/0/10     G1/0/11     G1/0/12
                         G1/0/13     G1/0/14     G1/0/15
```

G1/0/16	G1/0/17	G1/0/18
G1/0/19	G1/0/20	G1/0/21
G1/0/22	G1/0/23	G1/0/24
G1/0/25	G1/0/26	G1/0/27
G1/0/28	G1/0/29	G1/0/30
G1/0/31	G1/0/32	G1/0/33
G1/0/34	G1/0/35	G1/0/36
G1/0/37	G1/0/38	G1/0/39
G1/0/40	G1/0/41	G1/0/42
G1/0/43	G1/0/44	G1/0/45
G1/0/46	G1/0/47	G1/0/48
XG1/0/49	XG1/0/50	XG1/0/51
XG1/0/52		

[SW3]vlan 8   //在华三设备上手工创建 VLAN8
[SW3-vlan8]name QCNA   //在华三设备上提供"VLAN 命名"的命令
[SW3-vlan8]description ?
  TEXT Description string, 255 characters at most
[SW3-vlan8]description QCNA   //在华三设备上也提供"VLAN 描述"的命令,方便工程师识别这些 VLAN 用于什么场景。
[SW3]interface range GigabitEthernet 1/0/10 to GigabitEthernet 1/0/12   //使用"range"命令,使得在华三设备上一并配置 G0/0/10 到 G0/0/12 3 个端口
[SW3-if-range]port link-type access   //华三设备上默认为 access 模式,命令与华为设备上非常类似,但默认的端口类型不同
[SW3-if-range]port access vlan 8   //端口被静态地划分到 VLAN8,该命令与华为、思科设备上的不同,但相对而言更接近思科的命令
[SW3-if-range]exit
[SW3]display vlan brief   //查看 VLAN 创建和端口划分情况
Brief information about all VLANs:
Supported Minimum VLAN ID: 1
Supported Maximum VLAN ID: 4094
Default VLAN ID: 1

VLAN ID	Name	Port			
1	VLAN 0001	FG1/0/53	FG1/0/54	G1/0/1	
		G1/0/2	G1/0/3	G1/0/4	G1/0/5
		G1/0/6	G1/0/7	G1/0/8	G1/0/9
		G1/0/13	G1/0/14	G1/0/15	
		G1/0/16	G1/0/17	G1/0/18	
		G1/0/19	G1/0/20	G1/0/21	
		G1/0/22	G1/0/23	G1/0/24	
		G1/0/25	G1/0/26	G1/0/27	
		G1/0/28	G1/0/29	G1/0/30	
		G1/0/31	G1/0/32	G1/0/33	
		G1/0/34	G1/0/35	G1/0/36	
		G1/0/37	G1/0/38	G1/0/39	
		G1/0/40	G1/0/41	G1/0/42	
		G1/0/43	G1/0/44	G1/0/45	

		G1/0/46	G1/0/47	G1/0/48	
		XG1/0/49	XG1/0/50	XG1/0/51	
		XG1/0/52			
8	QCNA	G1/0/10	G1/0/11	G1/0/12	//在 VLAN8 中，已经增加了 G1/0/10 到 G1/0/12 端口

在 VLAN 划分完毕之后，进行终端设备的通信测试。

```
<H3C>ping 10.1.10.2    //测试三层数据通信，可以看到属于同一 VLAN 的终端已经可以实现通信，此处 PC1 已经与 PC2 通信
  ping 10.1.10.2 (10.1.10.2): 56 data bytes, press CTRL_C to break
  56 bytes from 10.1.10.2: icmp_seq=0 ttl=255 time=8.000 ms
  56 bytes from 10.1.10.2: icmp_seq=1 ttl=255 time=5.000 ms
  56 bytes from 10.1.10.2: icmp_seq=2 ttl=255 time=4.000 ms
  56 bytes from 10.1.10.2: icmp_seq=3 ttl=255 time=4.000 ms
  56 bytes from 10.1.10.2: icmp_seq=4 ttl=255 time=3.000 ms

  --- ping statistics for 10.1.10.2 ---
  5 packet(s) transmitted, 5 packet(s) received, 0.0% packet loss
  round-trip min/avg/max/std-dev = 3.000/4.800/8.000/1.720 ms
<H3C> ping 10.1.10.3    //PC1 已经与 PC3 通信
  ping 10.1.10.3 (10.1.10.3): 56 data bytes, press CTRL_C to break
  56 bytes from 10.1.10.3: icmp_seq=0 ttl=255 time=9.000 ms
  56 bytes from 10.1.10.3: icmp_seq=1 ttl=255 time=4.000 ms
  56 bytes from 10.1.10.3: icmp_seq=2 ttl=255 time=2.000 ms
  56 bytes from 10.1.10.3: icmp_seq=3 ttl=255 time=3.000 ms
  56 bytes from 10.1.10.3: icmp_seq=4 ttl=255 time=7.000 ms

  --- ping statistics for 10.1.10.3 ---
  5 packet(s) transmitted, 5 packet(s) received, 0.0% packet loss
  round-trip min/avg/max/std-dev = 2.000/5.000/9.000/2.608 ms
```

至此，华三设备基本的交换机 VLAN 接入实施完毕。

案例 10　交换机互连链路 VLAN 接入模式

交换机之间配置 VLAN 接入模式简书

作为一道经常在面试中出现的题，考官经常会这样问起：交换机之间能否实施接入模式呢？答案当然是可以的！当然这样做有它的优缺点。

交换机之间实施接入模式的优点：单个 VLAN 的流量可以独占某条链路，相对而言不容易产生流量拥塞，能够在 VLAN 较少的情况下保证较好的服务质量（QoS）。

交换机之间实施接入模式的缺点：由于每个 VLAN 都会"独享"一条链路，所以随着 VLAN 数量的增加，交换机之间占用的接口也随之增加，而通常情况下 VLAN 的数量众多，而一般的盒式交换机接口数为 24 个或 48 个（不包括光纤接口，光纤接口的数量非常少），交换机不能完全满足这种场景。

如果在交换机之间实施接入模式，属于同一 VLAN 的用户之间可以实现通信。当左侧用户发送数据到右侧用户时，当以太二型数据帧到达交换机时，在入方向强制增加 PVID green，则此时该数据帧携带 VLAN 标识（VLAN ID green），该数据帧可以发送到属于 VLAN green 的接口，在发送数据帧的同时会去掉 PVID green，这意味着此时发送的帧和用户初始发送的帧相同。

在数据帧到达右侧交换机时，处理过程和左侧交换机相同，先在入方向加入 PVID green 标识，然后再在出方向去掉 PVID green 标识，发送原始的以太二型数据帧到终端。

将交换机之间链路划分到同一个 VLAN 中如图 10-1 所示。在图 10-1 中，两台交换机的接口划入了同一 VLAN green。

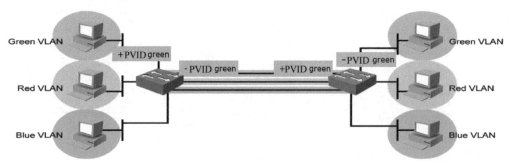

图 10-1　将交换机之间链路划分到同一个 VLAN 中

如果将交换机连接 PC 的接口划入不同 VLAN 中，PC 间可以通信吗？通常是不可以的。但是只要交换机将连接 PC 的接口和连接对端交换机的接口划入同一 VLAN 中，即使将交换机连接 PC 的接口划入不同 VLAN 中，PC 间依旧是可以通信的。将交换机之间链路划分到不同 VLAN 中如图 10-2 所示，图中 10.1.10.1 访问 10.1.10.2，在 SW1 上把数据帧发出时，会去掉 PVID "黑色 VLAN"，此时数据帧是原始以太网帧。当数据帧进入 SW2 时，会增加 SW2 设置的 VLAN ID 标识，即"绿色 VLAN"，而该帧自然可以在去掉 PVID "绿色 VLAN"标识后发送给 10.1.10.2。

图 10-2　将交换机之间链路划分到不同 VLAN 中

通过以下案例来验证。

10.1　交换机互连链路配置 VLAN 接入模式描述

在简书中，已经介绍了交换机互连链路 VLAN 接入模式的特定场景，接下来将给出具体描述。

R1 和 PC1 作为测试终端，把 SW1 的 2 个接口，以及 SW3 的 2 个接口划入 VLAN 8，以实现路由器和终端之间的通信（见图 10-3、图 10-4 和图 10-5）。

10.2　组网拓扑介绍

图 10-3、图 10-4 和图 10-5 分别展示了华为、思科和华三交换机之间实施接入模式，PC1（地址为 10.1.10.1/28）通过在两台交换机之间实施接入模式和 R1（地址为 10.1.10.14/28）实现通信。组网拓扑旨在告知读者交换机之间完全可以通过实施接入模式完成终端之间的通信。

10.3　配置要点

① 成功创建对应 VLAN。
② 同一交换机的多个接口必须被划分到同一 VLAN 中。
③ 确保终端的 IP 地址属于同一子网。

10.4　交换机互连链路 VLAN 接入模式配置详解

10.4.1　华为交换机互连链路 VLAN 接入模式配置

在华为交换机之间实施接入模式如图 10-3 所示。

为了不影响实验结果，须关闭接口 SW3 的接口 G0/0/1，否则可能会受到生成树协议的影响。为连接 PC 与路由器的所有接口配置接入模式，接入到同一 VLAN8。

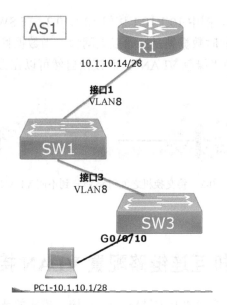

图 10-3 在华为交换机之间实施接入模式

```
[SW3]int g0/0/1
[SW3-GigabitEthernet0/0/1]shutdown
[SW3]int g0/0/3
[SW3-GigabitEthernet0/0/3]port link-type access
[SW3-GigabitEthernet0/0/3]port default vlan 8    //配置交换机互连接口的接入模式
!
[SW1]vlan 8
[SW1-vlan8]q
[SW1]interface g0/0/1
[SW1-GigabitEthernet0/0/1]port link-type access
[SW1-GigabitEthernet0/0/1]port default vlan 8    //连接路由器的接口配置接入模式,接入 VLAN8
[SW1-GigabitEthernet0/0/1]int g0/0/3
[SW1-GigabitEthernet0/0/3] port link-type access
[SW1-GigabitEthernet0/0/3] port default vlan 8   //交换机互连接口配置接入模式,接入 VLAN8,验
证 SW1 的配置结果
<SW1>display port vlan active
T=TAG U=UNTAG
--------------------------------------------------------------
Port              Link Type      PVID    VLAN List
--------------------------------------------------------------
G0/0/1            access         8       U: 8
G0/0/2            hybrid         1       U: 1
G0/0/3            access         8       U: 8
G0/0/4            hybrid         1       U: 1
G0/0/5            hybrid         1       U: 1
G0/0/6            hybrid         1       U: 1
G0/0/7            hybrid         1       U: 1
G0/0/8            hybrid         1       U: 1
```

G0/0/9	hybrid	1	U: 1
G0/0/10	hybrid	1	U: 1
G0/0/11	hybrid	1	U: 1
G0/0/12	hybrid	1	U: 1
G0/0/13	hybrid	1	U: 1
G0/0/14	hybrid	1	U: 1
G0/0/15	hybrid	1	U: 1
G0/0/16	hybrid	1	U: 1
G0/0/17	hybrid	1	U: 1
G0/0/18	hybrid	1	U: 1
G0/0/19	hybrid	1	U: 1
G0/0/20	hybrid	1	U: 1
G0/0/21	hybrid	1	U: 1
G0/0/22	hybrid	1	U: 1
G0/0/23	hybrid	1	U: 1
G0/0/24	hybrid	1	U: 1

在 R1 上配置与 PC 在同一网络的 IP 地址，然后进行测试。

```
[R1]interface g0/0/1
[R1-GigabitEthernet0/0/1]ip address 10.1.10.14 28
<R1>ping 10.1.10.1    //在交换机之间的配置没有任何问题，现在 R1 可以和终端进行通信。
  ping 10.1.10.1: 56   data bytes, press CTRL_C to break
    Reply from 10.1.10.1: bytes=56 Sequence=1 ttl=128 time=180 ms
    Reply from 10.1.10.1: bytes=56 Sequence=2 ttl=128 time=70 ms
    Reply from 10.1.10.1: bytes=56 Sequence=3 ttl=128 time=70 ms
    Reply from 10.1.10.1: bytes=56 Sequence=4 ttl=128 time=60 ms
    Reply from 10.1.10.1: bytes=56 Sequence=5 ttl=128 time=70 ms

  --- 10.1.10.1 ping statistics ---
    5 packet(s) transmitted
    5 packet(s) received
    0.00% packet loss
    round-trip min/avg/max = 60/90/180 ms
```

通过本例分析具体的数据通信过程如下：

① 当 PC1 向 R1 发送数据报文时，到达 SW3 的数据帧没有携带 TAG，则增加该接口的 PVID 8（添加默认的 VLAN TAG），那么该数据帧在交换机中以携带 TAG 8 的方式处理。

② SW3 只能将携带 TAG 8 的数据帧发送到相同的 VLAN 接口，则该数据帧从接口 3 发出时会去掉 TAG 8，此时为不携带 TAG 的数据帧。

③ 数据帧到达 SW1，由于不携带 TAG，所以按照步骤①处理，增加 PVID 8，数据帧开始携带 TAG 8。

④ SW1 重复步骤②，将数据帧从接口 1 发出去时，去掉 TAG 8，数据帧被发送到 R1。

R1 返回给 PC1 的数据帧重复步骤①到步骤④。

之前的实验在两台交换机之间的接入接口配置了相同的 VLAN，那么如果在另一台交换机互连的接口配置不同的 VLAN，数据如何通信呢？如下所示。

```
[SW1]vlan 10
[SW1-vlan10]int g0/0
```

```
[SW1-GigabitEthernet0/0/1]port link-type access
[SW1-GigabitEthernet0/0/1] port default vlan 10
[SW1-GigabitEthernet0/0/1]int g0/0/3
[SW1-GigabitEthernet0/0/3] port link-type access
[SW1-GigabitEthernet0/0/3] port default vlan 10    //SW1 的两个接口被划入 VLAN10，SW1 上连接
SW3 的接口被划入 VLAN10，而 SW1 连接 SW3 的 G0/0/3 接口被划入 VLAN8（即之前的配置不做改变）
!
[SW3]dis cu int g0/0/3
#
interface GigabitEthernet0/0/3
 port link-type access
 port default vlan 8
```

配置完成后，测试数据是否可以通信。

```
<R1>ping 10.1.10.2     //如您所见，此时 R1 和 PC1 之间依旧可以通信
  ping 10.1.10.2: 56   data bytes, press CTRL_C to break
    Reply from 10.1.10.2: bytes=56 Sequence=1 ttl=128 time=120 ms
    Reply from 10.1.10.2: bytes=56 Sequence=2 ttl=128 time=60 ms
    Reply from 10.1.10.2: bytes=56 Sequence=3 ttl=128 time=80 ms
    Reply from 10.1.10.2: bytes=56 Sequence=4 ttl=128 time=60 ms
    Reply from 10.1.10.2: bytes=56 Sequence=5 ttl=128 time=60 ms

  --- 10.1.10.2 ping statistics ---
    5 packet(s) transmitted
    5 packet(s) received
    0.00% packet loss
    round-trip min/avg/max = 60/76/120 ms
```

通过本例分析数据通信过程如下：

① 当 PC1 向 R1 发送数据报文时，到达 SW3 的数据帧没有 TAG，则增加该接口的 PVID 8（添加默认的 VLAN ID），那么在交换机中该数据帧被以携带 TAG 8 的方式处理。

② SW3 只能把携带 TAG 8 的数据帧发送到相同的 VLAN 接口，则当数据帧被从接口 3 发送出去时会去掉 TAG 8，此时为不携带 TAG 的数据帧。

③ 数据帧到达 SW1，由于不携带 TAG，所以按照步骤①处理，增加 PVID 10（添加默认的 VLAN ID），数据帧开始携带 TAG 10。

④ SW1 重复步骤②，数据帧在被从接口 1 发送出去时，去掉 TAG10，数据帧被发送到 R1。

R1 返回给 PC1 的数据帧重复步骤①到步骤④。问题的关键点在于当 SW3 从接口 3 发出去数据帧（或者 SW1 从接口 3 发出去帧）时，是不携带任何 TAG 的，交换机收到不携带的 TAG 的数据帧会增加配置的 VLAN ID。

当然，如果同一交换机上配置了不同的 VLAN，比如 SW3 的接口 G0/0/10 和接口 G0/0/3 配置接入到不同的 VLAN，那么数据是不可以通信的。

10.4.2　思科交换机互连链路 VLAN 接入模式配置

在思科交换机之间实施接入模式如图 10-4 所示。前面已经在 SW3 进行了部分 VLAN 的

划分实施,接下来请在之前实施的基础上继续完成本实验。

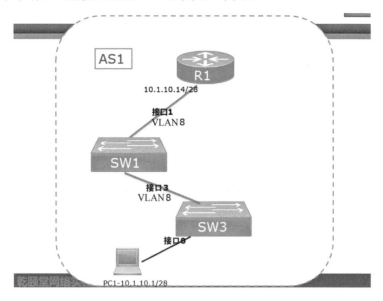

图 10-4 在思科交换机之间实施接入模式

```
SW1：
interface Ethernet0/1
  switchport access vlan 8
  switchport mode access
interface Ethernet0/3
  switchport access vlan 8
  switchport mode access
SW3：
interface Ethernet0/3
  switchport access vlan 8
  switchport mode access
interface Ethernet0/0
  switchport access vlan 8
  switchport mode access
```

SW3#show vlan brief   //验证 SW3 的 VLAN 简要信息

VLAN	Name	Status	Ports
1	default	active	E1/0, E1/1, E1/2, E1/3
8	QCNA	active	E0/0, E0/1, E0/2, E0/3 //接口 0 和接口 3 已

经被划分到 VLAN 8 中,那么只要这些终端属于同一子网就可以通信

1002	fddi-default	act/unsup	
1003	token-ring-default	act/unsup	
1004	fddinet-default	act/unsup	
1005	trnet-default	act/unsup	

请读者自行验证 SW1 的 VLAN 接入，注意查看交换机上接口 1 和接口 3 属于 VLAN 8。
请读者自行在 R1 上配置接口的地址，然后测试 3 层终端通信情况，如下所示。

```
R1-Cisco(config)#int e0/1
R1-Cisco(config-if)#ip address 10.1.10.14 255.255.255.240
PC1#ping 10.1.10.14
Type escape sequence to abort.
Sending 5, 100-byte ICMP Echos to 10.1.10.14, timeout is 2 seconds:
.!!!!
Success rate is 80 percent (4/5), round-trip min/avg/max = 1/2/4 ms
```

思科交换机设备之间采用接入模式实现终端通信案例完成。

10.4.3　华三交换机互连链路 VLAN 接入模式配置

在华三交换机之间实施接入模式如图 10-5 所示。

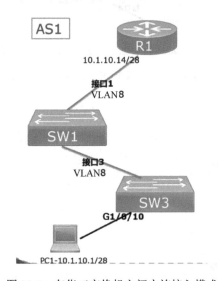

图 10-5　在华三交换机之间实施接入模式

前面已经在 SW3 上做了部分 VLAN 的划分实施，请在之前实施的基础上继续完成本实验（必须创建 VLAN 8），即把终端之间的所有接口都设置为 Access 模式。

```
H3C-SW1:
vlan 8 to 12
interface GigabitEthernet1/0/1
  port access vlan 8
interface GigabitEthernet1/0/3
  port access vlan 8
H3C-SW3:
interface GigabitEthernet1/0/3
  port link-mode bridge
  port access vlan 8
interface GigabitEthernet1/0/10
  port link-mode bridge
```

```
        port access vlan 8

[H3C-SW1]dis  vlan bri
8            VLAN 0008              G1/0/1         G1/0/3
[H3C-SW3]dis vlan bri
8            QCNA                   G1/0/3         G1/0/10
                                                   G1/0/11
```

请读者自行验证 SW1 的 VLAN 接入，注意查看交换机上接口 1 和接口 3 属于 VLAN8。

请读者自行在 R1 上配置接口的地址，然后测试 R1 和 PC1 的通信情况，如下所示，已经完成通信。

```
[H3C-R1-GigabitEthernet0/1]ping 10.1.10.1
56 bytes from 10.1.10.1: icmp_seq=0 ttl=255 time=11.000 ms
56 bytes from 10.1.10.1: icmp_seq=1 ttl=255 time=10.000 ms
56 bytes from 10.1.10.1: icmp_seq=2 ttl=255 time=5.000 ms
56 bytes from 10.1.10.1: icmp_seq=3 ttl=255 time=5.000 ms
56 bytes from 10.1.10.1: icmp_seq=4 ttl=255 time=5.000 ms
```

交换机之间实施接入模式可以用于同一网络两个终端之间独享交换机接口带宽的场景，在一定程度上可以保证该网络中数据在局域网中传输的服务质量。但这种场景仅是特例，在交换机之间经典的应用方式还是干道（Trunk）模式。将在下一个案例中进行学习。

至此，本案例实施完毕。

案例 11　交换机的 Trunk（干道）模式

交换机 Trunk 模式简书

Trunk 允许多个 VLAN 的流量在同一条链路上转发，Trunk 上可以转发多个 VLAN 的数据帧。它是一种"共享"模式。可以把 Trunk 想象为一条城际高速路，它总是拥有多条车道，城际高速路即干道链路，它可以在降低成本的同时提高转发效率。交换机之间每条 Trunk 链路可转发多个 VLAN 的数据，如图 11-1 所示，交换机之间使用 1 条链路就可以转发绿色、红色和蓝色 VLAN 的数据。

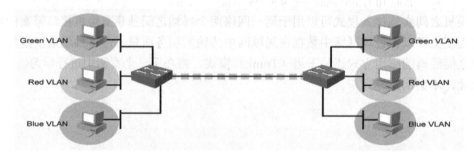

图 11-1　交换机之间使用 1 条 Trunk 链路转发多个 VLAN 的数据

Trunk 通过数据帧中的 TAG（TAG 中的 VLAN ID）指定 VLAN 流量的归属，一个真实的 TAG 封装如图 11-2 所示，这是一个实际的数据转发报文，可以看到增加了 TAG 的报文，图中的 VLAN ID 为 8。

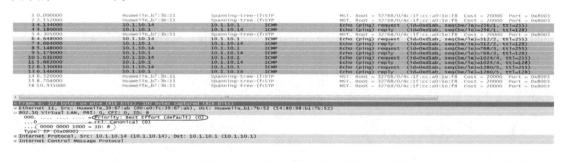

图 11-2　一个真实的 TAG 封装

图 11-3 所示为 Trunk 转发规则示意图，图中箭头的方向代表数据转发的方向。Trunk 转发规则如下。

① 在允许数据帧通过的前提下进行。
② 当 Trunk 端口收到数据帧时，如果该数据帧不包含 TAG，将打上端口的 PVID，如在

图 11-3 中的 ❶处增加了 PVID1；在 ❺处增加了 PVID20；如果该数据帧包含 TAG（VLAN20），则不改变任何标识，如图 11-3 中的 ❸处保留了 PVID1，而 ❼处保留了 TAG 标识 VLAN20。

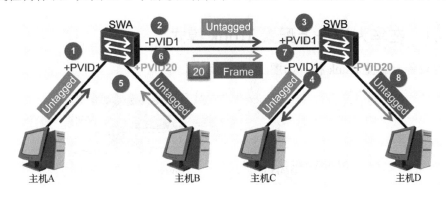

图 11-3　Trunk 转发规则示意图

③ 当 Trunk 端口发送数据帧时，该数据帧的 VLAN ID 在 Trunk 的允许发送列表中（即遵循规则：当数据帧的 TAG 与端口的 PVID 相同时，则剥离 TAG 发送，如图 11-3 中的 ❷处剥离掉 PVID1；当数据帧的 TAG 与端口的 PVID 不同时，则直接发送，如图 11-3 中的 ❻处发送的 TAG 依旧保留标识（VLAN20）。在图 11-3 中的 ❽处，原携带 TAG 20 的数据帧只能从此端口发送出去，同时去掉 TAG，将原始的、不带任何额外封装的以太二型数据帧发送给主机 D。

11.1　Trunk 模式应用场景

在交换机互连的所有端口上配置 Trunk 链路（注意，如果终端设备也支持 DOT1Q 封装，即可以承载多个 VLAN 的流量，则交换机连接终端的端口也可以使用 Trunk 封装，比如在支持虚拟化的服务器中，将连接交换机的端口配置为干道模式），用以实现在这些互连链路上承载多个 VLAN 的流量，使得接入层向汇聚层或核心层靠拢。此外，后面要介绍的单臂路由也会使用干道模式。

11.2　Trunk 模式配置案例拓扑说明

请参看图 11-4 和图 11-5 完成实施，我们会在 3 台交换机之间的互连链路上实施干道模式，用单条链路承载多个 VLAN 的数据，以保证设备之间的通信。

11.3　Trunk 模式配置要点

① 确定链路，正确配置 Trunk 命令。
② 查看和允许特定 VLAN 数据通过 Trunk 链路。
③ 正确配置链路两侧的 Trunk 参数。

11.4 Trunk 模式配置详解

11.4.1 在华为设备上配置 Trunk 模式

在华为设备上配置 Trunk 模式如图 11-4 所示。

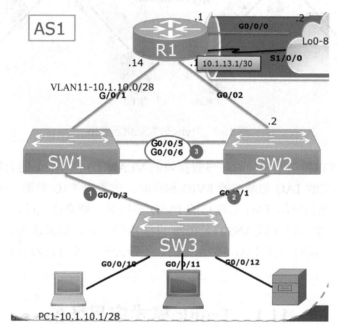

图 11-4　在华为设备上配置 Trunk 模式

使用图 11-4 中的 R1、SW1、SW3 以及 PC1 完成本案例，请在图中❶、❷、❸处配置 Trunk 模式。

本案例在上个案例的基础上实现，这是为了给读者演示如何在华为设备上修改链路类型。

```
[HW-SW1]vlan batch 8 9 10 11 12 99    //批量创建 VLAN，如果没有对应 VLAN，则交换机不转发
                                        该 VLAN 的数据帧，并在所有交换机上创建对应的 VLAN
[HW-SW2]vlan batch 8 9 10 11 12 99    //其他交换机也需要创建对应的 VLAN
[HW-SW3]vlan batch 8 9 10 11 12 99
[SW3]int g0/0/3
[SW3-GigabitEthernet0/0/3]dis th      //查看上个案例遗留的配置命令，依旧保持其接入模式
#
interface GigabitEthernet0/0/3
 port link-type access
 port default vlan 8
#
Return
[SW3-GigabitEthernet0/0/3]port link-type trunk   //在华为交换机上修改链路类型是比较麻烦的，如
                                                   此处所示，设备提示"请重新配置其到默认配置"
Error: Please renew the default configurations.
```

案例 11 交换机的 Trunk（干道）模式

[SW3-GigabitEthernet0/0/3]

为了能够修改接入端口为干道端口，请按照顺序完成如下配置：

[SW3-GigabitEthernet0/0/3]port default vlan 1   //先恢复端口默认的 VLAN1
Oct 10 2018 19:16:03-08:00 SW3 %%01IFNET/4/IF_STATE(l)[0]:Interface Vlanif1 has turned into UP state.
Oct 10 2018 19:16:03-08:00 SW3 %%01IFNET/4/LINK_STATE(l)[1]:The line protocol IP on the interface Vlanif1 has entered the UP state.
[SW3-GigabitEthernet0/0/3]port link-type trunk   //之后配置端口的链路类型为 Trunk 模式
[SW3-GigabitEthernet0/0/3]port trunk allow-pass vlan all   //华为设备上默认不允许任何 VLAN 的流量通过 Trunk 链路，需要手动配置允许的 VLAN 流量，此处配置为所有 VLAN
!
[SW1]int g0/0/3
[SW1-GigabitEthernet0/0/3]port link-type trunk
[SW1-GigabitEthernet0/0/3] port trunk allow-pass vlan 2 to 4094

注意，此时 R1 与 PC1 之间不能通信，因为根据之前的配置，SW1 的 G0/0/3 端口和 SW3 的 G0/0/10 端口并不属于同一个 VLAN，所以还需要做如下修改。

[SW1-GigabitEthernet0/0/3]int g0/0/1
[SW1-GigabitEthernet0/0/1]port default vlan 8
[SW1-GigabitEthernet0/0/1]dis th
#
interface GigabitEthernet0/0/1
 port link-type access
 port default vlan 8

交换机配置完毕，请在 R1 和 PC1 间进行数据通信测试。

```
<R1>ping 10.1.10.1
  ping 10.1.10.1: 56   data bytes, press CTRL_C to break
    Reply from 10.1.10.1: bytes=56 Sequence=1 ttl=128 time=110 ms
    Reply from 10.1.10.1: bytes=56 Sequence=2 ttl=128 time=60 ms
    Reply from 10.1.10.1: bytes=56 Sequence=3 ttl=128 time=70 ms
    Reply from 10.1.10.1: bytes=56 Sequence=4 ttl=128 time=50 ms
    Reply from 10.1.10.1: bytes=56 Sequence=5 ttl=128 time=70 ms

  --- 10.1.10.1 ping statistics ---
    5 packet(s) transmitted
    5 packet(s) received
    0.00% packet loss
  round-trip min/avg/max = 50/72/110 ms
```

数据从 R1 发往 PC1 的通信过程如下。

① R1 的 G0/0/1 端口在入方向上为原始数据帧增加了 PVID 8，此时数据帧在交换机内部携带 TAG 8。

② 当数据帧从 SW1 的 G0/0/3 端口发送出去时，首先通过查看发现该 Trunk 链路已经允许了 VLAN8 的流量通过，然后通过对比发现该数据帧的 VLAN ID(8)和该端口的 PVID（即默认为 1）不同，则数据帧保留 VLAN8 发送出去。

③ 数据帧到达 SW3 的 G0/0/3 端口时，该端口也允许 VLAN8 通过，同时该数据帧的

VLAN ID(8)不同于 PVID，即默认为 1，所以保留 TAG 8 并继续发送。

④ 携带 TAG 8 的数据帧可以从属于 VLAN8 的 SW3 的接入端口（G0/0/10 端口）发出去，此时剥离掉 TAG 8，将原始数据帧发送给客户端。

请读者自行完成❷和❸处的 Trunk 模式配置。

11.4.2 在思科设备上配置 Trunk 模式

在思科设备上配置 Trunk 模式如图 11-5 所示。

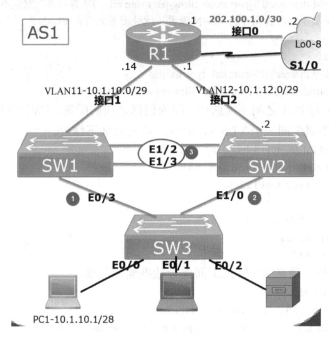

图 11-5 在思科设备上配置 Trunk 模式

本实验在 10.4.2 节配置的基础上完成（即在连接 PC 和路由器的端口配置接入模式），本例采用基本的交换机之间的 Trunk 命令，实验拓扑参看图 11-5，在图中❶、❷、❸处的链路上配置 Trunk 模式。

```
     SW3(config)#vlan 8,9,10,11,12,99    //思科设备批量创建 VLAN，VLAN 必须存在，否则不转发 VLAN
的数据帧，请在其他需要的交换机（SW1 和 SW2）上也创建 VLAN
     SW1(config)#vlan 8,9,10,11,12,99
     SW2(config)#vlan 8,9,10,11,12,99
     SW1&SW3：
     interface Ethernet0/3
       switchport trunk encapsulation dot1q    //使用 IEEE 802.1q 封装，部分思科设备使用默认的思科私
有的 ISL 封装，如果没有封装命令则代表仅支持 IEEE 802.1q 封装
       switchport mode trunk    //配置 Trunk 的"on"模式，即手动模式。思科设备上的 Trunk 链路默认
允许所有 VLAN 的流量通过，华为设备上的 Trunk 默认仅允许 VLAN1 的流量通过
     SW3#show interfaces trunk
     Port          Mode              Encapsulation    Status          Native vlan
     Et0/3         on                802.1q           trunking        1
```

Port	Vlans allowed on trunk
Et0/3	1-4094 //Trunk 上默认允许所有的 VLAN 的流量通过
Port	Vlans allowed and active in management domain
Et0/3	1,8-12,99
Port	Vlans in spanning tree forwarding state and not pruned
Et0/3	1,8-12,99

测试终端之间的数据通信。

```
PC1(config-if)#do ping 10.1.10.14
Type escape sequence to abort.
Sending 5, 100-byte ICMP Echos to 10.1.10.14, timeout is 2 seconds:
.!!!!
Success rate is 80 percent (4/5), round-trip min/avg/max = 1/1/2 ms
```

请读者自行完成❷和❸处的 Trunk 模式配置。

思科设备基本的 Trunk 模式配置完成。

11.4.3　在华三设备上配置 Trunk 模式

请读者参考 11.4.1 节内容将 G0/0/3 端口换成 G1/0/3 端口来完成相关配（本案例无图）。

在交换机互连的端口上配置 Trunk 模式，连接终端的端口依旧在 11.4.2 节的基础上接入到 VLAN8。

```
SW1&SW3
interface GigabitEthernet1/0/3
 port link-type trunk    //将端口的链路类型配置为 Trunk 模式
 port trunk permit vlan all    //Trunk 端口允许所有 VLAN 的流量通过，在现实网络中，用户可以根
据不同情况自行配置是否允许所有 VLAN 或特定 VLAN 数据通过 Trunk 端口
display port trunk //
Interface       PVID       VLAN Passing
G1/0/3           1          1, 8    //Trunk 端口允许默认的 VLAN1 和用户 VLAN8 的数据通过
```

测试用户的数据：

```
[H3C-R1]ping 10.1.10.1
ping 10.1.10.1 (10.1.10.1): 56 data bytes, press CTRL_C to break
56 bytes from 10.1.10.1: icmp_seq=0 ttl=255 time=8.000 ms
56 bytes from 10.1.10.1: icmp_seq=1 ttl=255 time=5.000 ms
56 bytes from 10.1.10.1: icmp_seq=2 ttl=255 time=5.000 ms
56 bytes from 10.1.10.1: icmp_seq=3 ttl=255 time=6.000 ms
56 bytes from 10.1.10.1: icmp_seq=4 ttl=255 time=5.000 ms
```

数据完成通信，本案例实施完毕。

案例 12 Trunk 上本征 VLAN 或 PVID 的最佳实践

Trunk 上的本征 VLAN 或 PVID 简书

几乎在所有的交换机上都存在一个默认 VLAN，即 VLAN1，这个 VLAN 非常特殊，该 VLAN 不允许执行删除操作，不允许修改名字，默认情况下所有端口都属于这个 VLAN。在 Trunk（干道）链路上存在一个 VLAN，当这个 VLAN 的流量在 Trunk 链路上通过时，不允许增加 TAG，这个 VLAN 即 VLAN1。思科称之为本征 VLAN（Native VLAN），华为称之为 Trunk 上的 PVID。

为什么会存在这样一个 VLAN 呢？读者可以查看图 12-1，本征 VLAN 的应用场景示意图，在该拓扑中有 3 个 VLAN（VLAN1、VLAN2、VLAN3），属于 VLAN1 的 PC①、PC②、PC③、PC④位于不同的交换机，而 PC②和 PC③由于设计或历史或资金原因，被一个 HUB（物理层设备）连接，当 PC①携带 TAG 的数据帧被从左侧交换机发出时，如果该数据帧依旧携带 TAG（即 VLAN1 的标识），它将被 HUB 直接转发给 PC②和 PC③，而 PC②和 PC③却不能识别 TAG，从而导致无法通信。这时本征 VLAN 或 PVID（后续这两者可能会混用或仅用其中一个名词）的作用就能体现出来了。本征 VLAN 的工作原理类似于 Access，默认的本征 VLAN（PVID）为 VLAN1，既然 VLAN1 是本征 VLAN，那么本来携带 TAG 的数据帧被从左侧交换机发送出去时会被去掉 TAG，此时的数据帧为原始的以太二型数据帧，将其发送给 PC②和 PC③时可以保证通信。

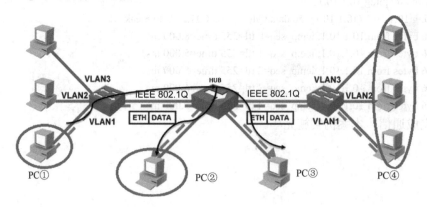

图 12-1 本征 VLAN 的应用场景示意图

本征 VLAN 的数据帧在 Trunk 链路上不带标识，交换机收到不带标识的帧将会转发到本征 VLAN。

如果一条Trunk链路上两侧的本征VLAN不同,而本征VLAN类似于接入端口的VLAN,会产生跨越VLAN攻击现象,引发VLAN访问的混乱。

为了解决这种情况,有以下两种基本解决方案。
- 主动解决方案：在Trunk上使本征VLAN也携带TAG,华为、H3C的Trunk配置没有该方案（混杂端口有类似功能）；
- 被动解决方案：在Trunk上使用一个没有业务的VLAN作为本征VLAN,使得原来的VLAN业务不受影响。

12.1 调整Trunk上的本征VLAN或PVID

调整Trunk上的本征VLAN或PVID可以使属于不同VLAN的设备实现通信,所以进一步的解决方案是如何避免混乱的VLAN业务互访。

12.2 实施拓扑说明

华为设备Trunk上的PVID VLAN示意图如图12-2所示,PC1接入SW3的端口被划分到VLAN10中,同时将SW3的Trunk端口的PVID设置为VLAN10；R1接入SW1的端口被划分到VLAN8中,同时将SW1的Trunk端口的PVID设置为VLAN8。

图12-2 华为设备Trunk上的PVID VLAN示意图

12.3 调整Trunk上本征VLAN或PVID的配置要点

① 创建对应的VLAN。
② 配置交换机之间的Trunk模式。
③ 修改同一交换机上Trunk端口的本征VLAN,使其等同于接入端口的VLAN。
④ 终端设备属于同一子网。

12.4 调整Trunk上本征VLAN或PVID的配置详解

12.4.1 在华为设备的Trunk链路上调整PVID VLAN

在11.4.1节的基础上,在两个交换机上完成如下的配置。
SW1：
```
interface GigabitEthernet0/0/1
 port link-type access
 port default vlan 8
```

```
        description TOR1
        #
        interface GigabitEthernet0/0/3
          port link-type trunk
          port trunk pvid vlan 8    //Trunk 上的 PVID 默认为 1，现在被修改为 8，该 VLAN 等于交换机上
Access 端口的 VLAN，即 VLAN8
          port trunk allow-pass vlan 2 to 4094
          description TOSW3
        !
```

SW3：

```
        interface GigabitEthernet0/0/3
          port link-type trunk
          port trunk pvid vlan 10    //Trunk 上的 PVID 默认为 1，现在被修改为 10，该 VLAN 等于交换机上
Access 端口的 VLAN，即 VLAN10
          port trunk allow-pass vlan 2 to 4094
          description TOSW1
        #
        interface GigabitEthernet0/0/10
          port link-type access
          port default vlan 10
          description TOPC1
```

验证实施结果如下。

```
[SW1]display port vlan active
T=TAG U=UNTAG
--------------------------------------------------------
Port              Link Type    PVID    VLAN List
--------------------------------------------------------
G0/0/1            access       8       U: 8
G0/0/2            hybrid       1       U: 1
G0/0/3            trunk        8       U: 8
                                       T: 1 10 20 99

[SW3]dis port vlan active
T=TAG U=UNTAG
--------------------------------------------------------
Port              Link Type    PVID    VLAN List
--------------------------------------------------------
G0/0/1            hybrid       1       U: 1
G0/0/2            hybrid       1       U: 1
G0/0/3            trunk        10      U: 10
                                       T: 1 8 20 99
G0/0/4            hybrid       1       U: 1
G0/0/5            hybrid       1       U: 1
G0/0/6            hybrid       1       U: 1
G0/0/7            hybrid       1       U: 1
G0/0/8            hybrid       1       U: 1
G0/0/9            hybrid       1       U: 1
G0/0/10           access       10      U: 10
```

案例 12　Trunk 上本征 VLAN 或 PVID 的最佳实践

对 R1 到 PC1 的数据进行测试会发现，接入 VLAN8 的用户可以与接入 VLAN10 的用户通信，这在一定程度上是具备危害性的，我们称之为跨越 VLAN 攻击。测试代码如下所示。

```
<R1>ping 10.1.10.1
   ping 10.1.10.1: 56   data bytes, press CTRL_C to break
     Reply from 10.1.10.1: bytes=56 Sequence=1 ttl=128 time=50 ms
     Reply from 10.1.10.1: bytes=56 Sequence=2 ttl=128 time=50 ms
     Reply from 10.1.10.1: bytes=56 Sequence=3 ttl=128 time=70 ms
     Reply from 10.1.10.1: bytes=56 Sequence=4 ttl=128 time=60 ms
     Reply from 10.1.10.1: bytes=56 Sequence=5 ttl=128 time=50 ms

   --- 10.1.10.1 ping statistics ---
     5 packet(s) transmitted
     5 packet(s) received
     0.00% packet loss
     round-trip min/avg/max = 50/56/70 ms
```

在此通信过程中，数据帧在 Trunk 上通信时不携带任何的 TAG，该过程类似于 Access 方式。Trunk 上不携带 VLAN TAG 的数据帧如图 12-3 所示。

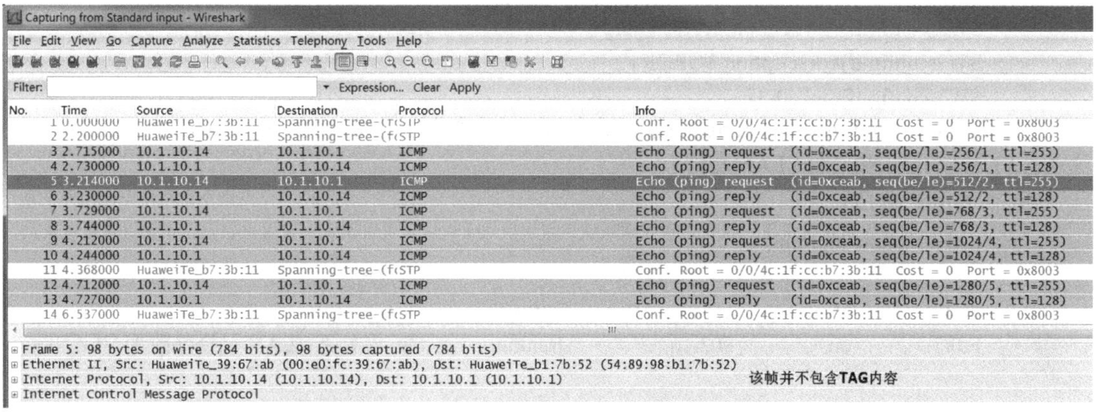

图 12-3　Trunk 上不携带 VLAN TAG 的数据帧

若 PVID 在 Trunk 上携带 TAG 则可以解决跨越 VLAN 攻击的问题，但是华为设备上并没有这种方案（后续在介绍 Hybrid 端口时我们再讨论）。

此处的最佳解决方案如下：在 Trunk 上把 PVID 修改为一个没有业务并且是非 VLAN1 的 VLAN。请读者在案例 12 实施完毕之后修改所有互连端口为 Trunk 端口，将 PVID 修改为 VLAN 99（请创建对应 VLAN）。

```
interface GigabitEthernet0/0/3
 port link-type trunk
 port trunk pvid vlan 99    //将 PVID 修改为一个没有业务的 VLAN，这样就不会有业务数据被攻击
 port trunk allow-pass vlan 2 to 4094
```

12.4.2　在思科设备的 Trunk 链路上调整本征 VLAN

在 11.4.2 节的基础上，在两个交换机 SW1 和 SW2 上完成如下配置。

```
C-SW1(config)#do sh run int e0/1
```

```
Building configuration...

Current configuration : 79 bytes
!
interface Ethernet0/1
 switchport access vlan 8
 switchport mode access
 description TOR1
end


C-SW1(config)#int e0/3
C-SW1(config-if)#switchport trunk native vlan 8    //将交换机互连端口的本征 VLAN 调整为 VLAN8，
注意该 VLAN 和接入端口的 VLAN 相同
*Nov 21 11:38:13.117: %CDP-4-NATIVE_VLAN_MISMATCH: Native VLAN mismatch discovered
on Ethernet0/3 (8), with SW3 Ethernet0/3 (10).   //调整完本征 VLAN 之后，E0/3 端口的本征 VLAN 为 VLAN8，
而对端为 VLAN 10，此时 CDP 协议会报错，但并不影响通信。在真机上还需要关闭对应 VLAN 的生成树
!
SW3(config-if)#int e0/0
SW3(config-if)#switchport access vlan 10
SW3(config-if)#description TOPC1
SW3(config-if)#int e0/3
SW3(config-if)#switchport trunk native vlan 10    //将交换机互连端口的本征 VLAN 调整为 VLAN10，
注意该 VLAN 和接入端口的 VLAN 相同
```

验证 Trunk 端口：

```
C-SW1#show int trunk

Port        Mode        Encapsulation     Status        Native vlan
Et0/3       on          802.1q            trunking      8     //本征 VLAN 已经修改为 VLAN8

Port        Vlans allowed on trunk
Et0/3       1-4094

Port        Vlans allowed and active in management domain
Et0/3       1,8-12,99

Port        Vlans in spanning tree forwarding state and not pruned
Et0/3       1,8-12,99
SW3#show int trunk

Port        Mode        Encapsulation     Status        Native vlan
Et0/3       on          802.1q            trunking      10

Port        Vlans allowed on trunk
Et0/3       1-4094

Port        Vlans allowed and active in management domain
Et0/3       1,8-12,99
```

```
Port          Vlans in spanning tree forwarding state and not pruned
Et0/3         1,8-12,99
```

在配置完毕接入端口和 Trunk 端口后，测试通信情况：

```
R1-Cisco#ping 10.1.10.1    //隶属于 VLAN8 和 VLAN10 的终端完成了通信
Sending 5, 100-byte ICMP Echos to 10.1.10.1, timeout is 2 seconds:
!!!!!
```

思科针对本征 VLAN 的问题有以下两种解决方案。

① 主动方案：既然问题是由于本征 VLAN 不携带 TAG 导致的，那么只要使本征 VLAN 携带 TAG 即可。在思科设备上执行以下命令。

```
C-SW1(config)#vlan dot1q tag native    //与大部分模拟器不同，笔者共享的模拟器可以支持该命令，在执行该命令后，前面实验中的终端将不能通信，这说明操作成功
```

② 被动方案：将本征 VLAN 修改为一个特定的 VLAN 或者没有业务的 VLAN，而不使用 VLAN1 作为本征 VLAN，比如改为 VLAN99，执行命令如下：

```
C-SW1(config-if)#switchport trunk native vlan 99    //VLAN 99 中没有实际的业务，同时其流量又不属于 VLAN1，避免默认的 VLAN1 造成的混乱
```

为了方便后续实验，请读者自行在交换机之间互连的端口实施 Trunk 模式，并修改本征 VLAN 为 VLAN99。

本案例实施完毕。

案例 13　华为交换机上的 Hybrid（混杂）端口

Hybrid 端口模式简书

Hybrid 端口既可以用于连接不能识别 TAG 的用户终端（如用户主机、服务器等）和网络设备（如 HUB），又可以用于连接交换机、路由器，以及可同时收发 Tagged 数据帧和 Untagged 数据帧的语音终端、AP。它可以允许多个 VLAN 的带 TAG 的数据帧通过，且允许从该类端口发出的数据帧根据需要进行配置，使某些 VLAN 的数据帧携带 TAG（即不剥除 TAG）、某些 VLAN 的数据帧不携带 TAG（即剥除 TAG）。

Hybrid 端口和 Trunk 端口在很多应用场景下可以通用，但在某些应用场景下，只能使用 Hybrid 端口。比如在一个端口连接不同 VLAN 网段的场景中，因为一个端口需要给多个 Untagged 报文添加 TAG，所以必须使用 Hybrid 端口（这点类似于案例 12 中思科设备为本征 VLAN 添加 TAG）。

当使用 Hybrid 端口连接交换机时，推荐使用 TAG 来处理多个 VLAN 的流量（该功能等同于 Trunk）。

该类型的显著特点：连接终端时，入方向可以增加一个 PVID（类似于接入模式），在出方向允许多个 VLAN 的数据帧转发出去并移除 PVID（不同于接入模式的单个 VLAN）。

Hybrid 端口接收数据帧处理流程如图 13-1 所示。

Hybrid 端口发送数据帧处理流程如图 13-2 所示。

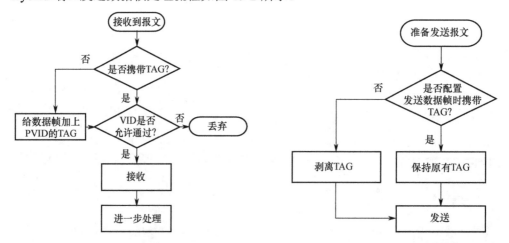

图 13-1　Hybrid 端口接收数据帧处理流程　　图 13-2　Hybrid 端口发送数据帧处理流程

13.1 交换机混杂模式应用场景

混杂模式（Hybrid）结合了 Access 和 Trunk 的特点，它的特点是可以灵活地操控 TAG 和 PVID，在连接终端端口的入方向，会对以太数据帧增加 1 个 PVID，在连接终端端口的出方向可以在允许多个 VLAN 的流量转发出去的同时剥离掉 VLAN 标识。在本例中将通过控制链路上的 VLAN ID 灵活地操控数据访问。

13.2 交换机混杂模式配置拓扑说明

图 13-3 为 Hybrid 端口实施示意图，参照图 13-3 所示拓扑的地址和 VLAN 设计，在两个交换机上实施 VLAN 的混杂模式，使得 PC1（10.1.10.1）和 PC2（10.1.10.2）都可以和路由器（10.1.10.14）通信，但是 PC1 和 PC2 之间不能通信。

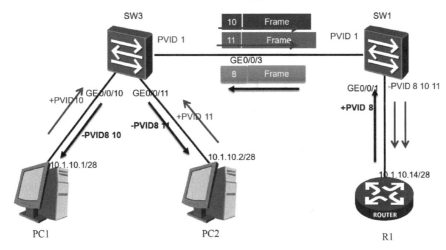

图 13-3 Hybrid 端口实施示意图

13.3 混杂模式配置要点

① 把原有的端口模式修改为混杂模式。
② 配置交换机互连端口，以类似 Trunk 的方式处理 TAG。
③ 配置交换机连接终端的端口，在入方向设置一个 PVID，在出方向设置多个 PVID。

13.4 混杂模式配置步骤详解

13.4.1 将交换机互连端口配置为混杂模式

由于交换机互连端口通常允许多个 VLAN 的流量携带 TAG 通过，所以可以采用 Hybrid 模式配置交换机互连端口。

SW1:

  [HW-SW1-GigabitEthernet0/0/3]port trunk pvid vlan 1　　//将端口类型从其他模式修改为混杂模式，并恢复默认参数

  [HW-SW1-GigabitEthernet0/0/3]undo port trunk allow-pass vlan 2 to 4094　　//将端口类型从其他模式修改为混杂模式，并恢复默认参数

  [HW-SW1-GigabitEthernet0/0/3]port link-type hybrid

  [HW-SW1-GigabitEthernet0/0/3]port hybrid tagged vlan all　　//类似于Trunk，连接其他交换机，混杂端口此时允许所有的数据帧携带TAG通过，包含VLAN1

SW3：

连接交换机的端口。

  [HW-SW3-GigabitEthernet0/0/3]port trunk pvid vlan 1
  [HW-SW3-GigabitEthernet0/0/3]undo port trunk allow-pass vlan 2 to 4094
  [HW-SW3-GigabitEthernet0/0/3]port trunk pvid vlan 1
  [HW-SW3-GigabitEthernet0/0/3]port link-type hybrid
  [HW-SW3-GigabitEthernet0/0/3]port hybrid tagged vlan all

验证配置结果（请注意，仅仅验证配置是"小白"的表现，要通过验证命令而非验证配置来判断问题）。

```
[SW3]display port vlan active
T=TAG U=UNTAG
--------------------------------------------------------------
Port             Link Type      PVID     VLAN List
--------------------------------------------------------------
GE0/0/1          hybrid         1        U: 1
GE0/0/2          hybrid         1        U: 1
GE0/0/3          hybrid         1        T: 1 8 to 12 99
GE0/0/4          hybrid         1        U: 1
GE0/0/5          hybrid         1        U: 1
GE0/0/6          hybrid         1        U: 1
GE0/0/7          hybrid         1        U: 1
GE0/0/8          hybrid         1        U: 1
GE0/0/9          hybrid         1        U: 1
GE0/0/10         hybrid         1        U: 1
GE0/0/11         hybrid         1        U: 1
GE0/0/12         hybrid         1        U: 1
GE0/0/13         hybrid         1        U: 1
GE0/0/14         hybrid         1        U: 1
GE0/0/15         hybrid         1        U: 1
GE0/0/16         hybrid         1        U: 1
GE0/0/17         hybrid         1        U: 1
GE0/0/18         hybrid         1        U: 1
GE0/0/19         hybrid         1        U: 1
GE0/0/20         hybrid         1        U: 1
GE0/0/21         hybrid         1        U: 1
GE0/0/22         hybrid         1        U: 1
GE0/0/23         hybrid         1        U: 1
GE0/0/24         hybrid         1        U: 1
[SW1-GigabitEthernet0/0/3]display port vlan active
T=TAG U=UNTAG
--------------------------------------------------------------
```

案例 13　华为交换机上的 Hybrid（混杂）端口

Port	Link Type	PVID	VLAN List
GE0/0/1	hybrid	1	U: 1
GE0/0/2	hybrid	1	U: 1
GE0/0/3	hybrid	1	T: 1 8 to 12 99
GE0/0/4	hybrid	1	U: 1
GE0/0/5	hybrid	1	U: 1
GE0/0/6	hybrid	1	U: 1
GE0/0/7	hybrid	1	U: 1
GE0/0/8	hybrid	1	U: 1
GE0/0/9	hybrid	1	U: 1
GE0/0/10	hybrid	1	U: 1
GE0/0/11	hybrid	1	U: 1
GE0/0/12	hybrid	1	U: 1
GE0/0/13	hybrid	1	U: 1
GE0/0/14	hybrid	1	U: 1
GE0/0/15	hybrid	1	U: 1
GE0/0/16	hybrid	1	U: 1
GE0/0/17	hybrid	1	U: 1
GE0/0/18	hybrid	1	U: 1
GE0/0/19	hybrid	1	U: 1
GE0/0/20	hybrid	1	U: 1
GE0/0/21	hybrid	1	U: 1
GE0/0/22	hybrid	1	U: 1
GE0/0/23	hybrid	1	U: 1
GE0/0/24	hybrid	1	U: 1

将 GE0/0/3 端口上允许的 VLAN 的 PVID 设置为 VLAN99，而不使用默认的 VLAN1。

　　[Sw1-GigabitEthernet0/0/3] port hybrid pvid vlan 99　　//设置该端口的 PVID 为 VLAN99 而不使用默认的 VLAN1

　　[Sw3-GigabitEthernet0/0/3] port hybrid pvid vlan 99

验证配置结果：

　　[SW1-GigabitEthernet0/0/3]display port vlan active
　　T=TAG U=UNTAG

Port	Link Type	PVID	VLAN List
GE0/0/1	hybrid	1	U: 1
GE0/0/2	hybrid	1	U: 1
GE0/0/3	hybrid	99	T: 1 8 to 12 99 //允许所有的 VLAN 通过交换机互连链路，PVID 所在 VLAN 为 99。VLAN 99 的数据帧在通过该链路时也携带 TAG 标识
GE0/0/4	hybrid	1	U: 1
GE0/0/5	hybrid	1	U: 1
GE0/0/6	hybrid	1	U: 1
GE0/0/7	hybrid	1	U: 1
GE0/0/8	hybrid	1	U: 1
GE0/0/9	hybrid	1	U: 1
GE0/0/10	hybrid	1	U: 1
GE0/0/11	hybrid	1	U: 1

GE0/0/12	hybrid	1	U: 1
GE0/0/13	hybrid	1	U: 1
GE0/0/14	hybrid	1	U: 1
GE0/0/15	hybrid	1	U: 1
GE0/0/16	hybrid	1	U: 1
GE0/0/17	hybrid	1	U: 1
GE0/0/18	hybrid	1	U: 1
GE0/0/19	hybrid	1	U: 1
GE0/0/20	hybrid	1	U: 1
GE0/0/21	hybrid	1	U: 1
GE0/0/22	hybrid	1	U: 1
GE0/0/23	hybrid	1	U: 1
GE0/0/24	hybrid	1	U: 1

接下来配置连接终端的端口。

13.4.2 将连接终端的端口配置为混杂模式

需要修改连接终端的入方向和出方向 VLAN 以满足要求，具体如下。

```
[SW1]int g0/0/1
[SW1-GigabitEthernet0/0/1]port hybrid pvid vlan 8    //在连接路由器的混杂端口入方向增加 PVID 8
[SW1-GigabitEthernet0/0/1]port hybrid untagged vlan 8 10 11    //注意该命令，接入端口仅允许一个
```
相同的 VLAN 的数据在出方向通过，而混杂端口在出方向允许多个 VLAN 的数据通过，尤其是 VLAN10 和 VLAN11 的标识其实是 PC1 和 PC2 发送的数据在到达 SW3 时增加的 TAG，所以这些数据才可以发送到 R1 ！

```
[SW3-GigabitEthernet0/0/3]int g0/0/10
[SW3-GigabitEthernet0/0/10]port hybrid pvid vlan 10    //在连接 PC1 的混杂端口入方向增加 PVID 10
[SW3-GigabitEthernet0/0/10]port hybrid untagged vlan 8 10    //连接 PC1 的混杂端口出方向允许
```
VLAN8 和 VLAN11 的数据通过。注意 VLAN8 的标识是 R1 发送数据时增加的 TAG，但没有允许 VLAN10 的数据通过，所以 PC1 和 PC2 不可通信，而 R1 可以和 PC1 通信

```
[SW3-GigabitEthernet0/0/10]int g0/0/11
[SW3-GigabitEthernet0/0/11]port hybrid pvid vlan 11
[SW3-GigabitEthernet0/0/11]port hybrid untagged vlan 8 11
```

验证配置结果。

```
[SW1-GigabitEthernet0/0/1]display port vlan active
T=TAG U=UNTAG
```

Port	Link Type	PVID	VLAN List
GE0/0/1	hybrid	8	U: 1 8 10 to 11
GE0/0/2	hybrid	1	U: 1
GE0/0/3	hybrid	99	T: 1 8 to 12 99
GE0/0/4	hybrid	1	U: 1
GE0/0/5	hybrid	1	U: 1
GE0/0/6	hybrid	1	U: 1
GE0/0/7	hybrid	1	U: 1
GE0/0/8	hybrid	1	U: 1
GE0/0/9	hybrid	1	U: 1
GE0/0/10	hybrid	1	U: 1
GE0/0/11	hybrid	1	U: 1

Port	Link Type	PVID	VLAN List
GE0/0/12	hybrid	1	U: 1
GE0/0/13	hybrid	1	U: 1
GE0/0/14	hybrid	1	U: 1
GE0/0/15	hybrid	1	U: 1
GE0/0/16	hybrid	1	U: 1
GE0/0/17	hybrid	1	U: 1
GE0/0/18	hybrid	1	U: 1
GE0/0/19	hybrid	1	U: 1
GE0/0/20	hybrid	1	U: 1
GE0/0/21	hybrid	1	U: 1
GE0/0/22	hybrid	1	U: 1
GE0/0/23	hybrid	1	U: 1
GE0/0/24	hybrid	1	U: 1

[SW3-GigabitEthernet0/0/11]display port vlan active
T=TAG U=UNTAG

Port	Link Type	PVID	VLAN List
GE0/0/1	hybrid	1	U: 1
GE0/0/2	hybrid	1	U: 1
GE0/0/3	hybrid	99	T: 1 8 to 12 99
GE0/0/4	hybrid	1	U: 1
GE0/0/5	hybrid	1	U: 1
GE0/0/6	hybrid	1	U: 1
GE0/0/7	hybrid	1	U: 1
GE0/0/8	hybrid	1	U: 1
GE0/0/9	hybrid	1	U: 1
GE0/0/10	hybrid	10	U: 1 8 10
GE0/0/11	hybrid	11	U: 1 8 11
GE0/0/12	hybrid	1	U: 1
GE0/0/13	hybrid	1	U: 1
GE0/0/14	hybrid	1	U: 1
GE0/0/15	hybrid	1	U: 1
GE0/0/16	hybrid	1	U: 1
GE0/0/17	hybrid	1	U: 1
GE0/0/18	hybrid	1	U: 1
GE0/0/19	hybrid	1	U: 1
GE0/0/20	hybrid	1	U: 1
GE0/0/21	hybrid	1	U: 1
GE0/0/22	hybrid	1	U: 1
GE0/0/23	hybrid	1	U: 1
GE0/0/24	hybrid	1	U: 1

在配置完毕之后，发送数据进行测试。
测试 R1 到 PC1 的流量：

```
<R1>ping 10.1.10.1
  ping 10.1.10.1: 56   data bytes, press CTRL_C to break
    Reply from 10.1.10.1: bytes=56 Sequence=1 ttl=128 time=70 ms
    Reply from 10.1.10.1: bytes=56 Sequence=2 ttl=128 time=90 ms
    Reply from 10.1.10.1: bytes=56 Sequence=3 ttl=128 time=70 ms
    Reply from 10.1.10.1: bytes=56 Sequence=4 ttl=128 time=80 ms
```

```
        Reply from 10.1.10.1: bytes=56 Sequence=5 ttl=128 time=70 ms

    --- 10.1.10.1 ping statistics ---
      5 packet(s) transmitted
      5 packet(s) received
      0.00% packet loss
      round-trip min/avg/max = 70/76/90 ms
```

当然，R1 也可以和 PC2 通信。

```
    <R1>ping 10.1.10.2
      ping 10.1.10.2: 56    data bytes, press CTRL_C to break
        Reply from 10.1.10.2: bytes=56 Sequence=1 ttl=128 time=80 ms
        Reply from 10.1.10.2: bytes=56 Sequence=2 ttl=128 time=60 ms
        Reply from 10.1.10.2: bytes=56 Sequence=3 ttl=128 time=70 ms
        Reply from 10.1.10.2: bytes=56 Sequence=4 ttl=128 time=80 ms
        Reply from 10.1.10.2: bytes=56 Sequence=5 ttl=128 time=70 ms

    --- 10.1.10.2 ping statistics ---
      5 packet(s) transmitted
      5 packet(s) received
      0.00% packet loss
      round-trip min/avg/max = 60/72/80 ms
```

此时 PC1 和 PC2 无法通信（见图 13-4），因为当 PC1 把数据发送给 PC2 时，在 SW3 的 GE0/0/10 端口增加了 PVID 10，此时数据帧携带了 TAG 10，但 GE0/0/11 不允许该帧从本端口发送出去（允许 VLAN8 和 VLAN11 的数据通过，但没有 VLAN10 的数据通过）。

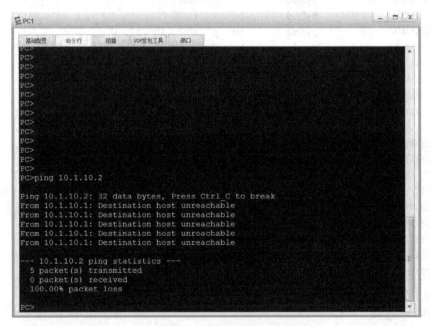

图 13-4 PC1 和 PC2 无法通信

除非在 GE0/0/11 端口以 UNTAG 的方式增加 VLAN10，当然 GE0/0/10 端口也需要以 UNTAG 的方式增加 VLAN11，配置如下：

```
[SW3-GigabitEthernet0/0/10]port hybrid untagged vlan 8 10 11
[SW3-GigabitEthernet0/0/10]int g0/0/11
[SW3-GigabitEthernet0/0/11]port hybrid untagged vlan 8 10 11
```

验证配置结果：

```
[SW3-GigabitEthernet0/0/10]display port vlan active
T=TAG U=UNTAG
-------------------------------------------------------------
Port            Link Type    PVID    VLAN List
-------------------------------------------------------------
GE0/0/1         hybrid       1       U: 1
GE0/0/2         hybrid       1       U: 1
GE0/0/3         hybrid       99      T: 1 8 to 12 99
GE0/0/4         hybrid       1       U: 1
GE0/0/5         hybrid       1       U: 1
GE0/0/6         hybrid       1       U: 1
GE0/0/7         hybrid       1       U: 1
GE0/0/8         hybrid       1       U: 1
GE0/0/9         hybrid       1       U: 1
GE0/0/10        hybrid       10      U: 1 8 10 to 11
GE0/0/11        hybrid       11      U: 1 8 10 to 11
GE0/0/12        hybrid       1       U: 1
GE0/0/13        hybrid       1       U: 1
GE0/0/14        hybrid       1       U: 1
GE0/0/15        hybrid       1       U: 1
GE0/0/16        hybrid       1       U: 1
GE0/0/17        hybrid       1       U: 1
GE0/0/18        hybrid       1       U: 1
GE0/0/19        hybrid       1       U: 1
GE0/0/20        hybrid       1       U: 1
GE0/0/21        hybrid       1       U: 1
GE0/0/22        hybrid       1       U: 1
GE0/0/23        hybrid       1       U: 1
GE0/0/24        hybrid       1       U: 1
```

如图 13-5 所示，验证 PC1 和 PC2 的数据通信情况，PC1 和 PC2 可以通信。

到此 Hybrid 的链路类型实施完毕，后期在 NP 阶段我们还会学习华为的 MUX VLAN 和端口保护技术，也能起到类似的控制 VLAN 间数据通信的作用。一般情况下由于 Hybrid 对 VLAN 的控制过于灵活，所以只会用在某些特定环境。

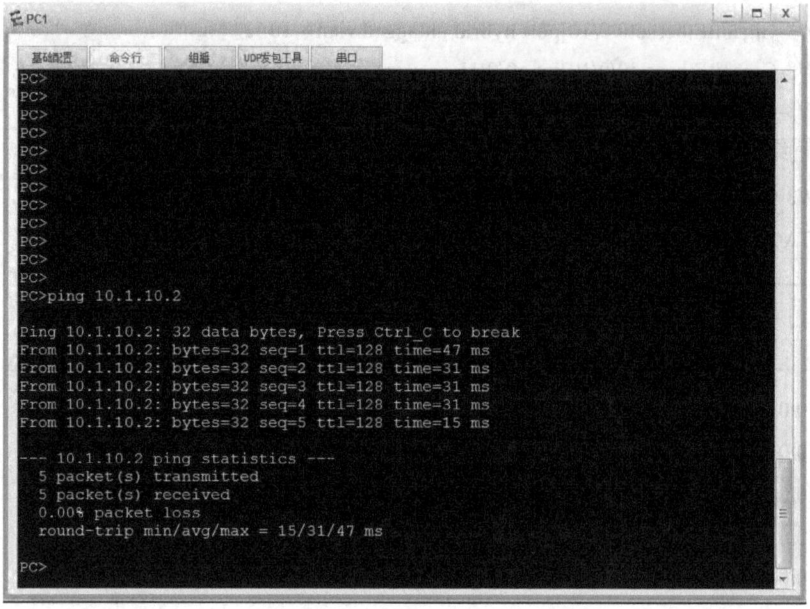

图 13-5　PC1 和 PC2 可以通信

在华三设备上配置混杂端口的基本命令如表 13-1 所示。

表 13-1　在华三设备上配置混杂端口的基本命令

配置端口的链路类型为 Hybrid 类型（默认 Access）
[Switch-Ethernet1/0/1] port link-type hybrid
允许指定的 VLAN 的数据通过当前 Hybrid 端口
[Switch-Ethernet1/0/1] port hybrid vlan vlan-id-list { tagged	untagged }
设置 Hybrid 端口的默认 VLAN
[Switch-Ethernet1/0/1] port hybrid pvid vlan vlan-id

至此，本案例实施完毕。

案例14 华为华三的标准生成树协议

生成树协议简书

生成树协议（Spanning Tree Protocol，STP）用于解决在二层交换网络中由于部署冗余链路而造成的转发环路问题。在二层网络中，一旦存在环路就会造成数据帧在环路内不断循环和增生，产生广播风暴而导致网络瘫痪。在这种场景下，生成树协议应运而生，生成树协议是一种二层管理协议，它通过有选择性地阻塞网络冗余链路来达到消除网络二层环路的目的，同时具备链路的备份功能。生成树协议和其他协议一样，是随着网络的不断发展而不断更新换代的。最初被广泛应用的是 IEEE 802.1d-1998 STP，并以此为基础产生了 2.0 版本和 3.0 版本，分别是 IEEE 802.1w RSTP（Rapid Spanning Tree Protocol，快速生成树协议）和 IEEE 802.1s MSTP（Multiple Spanning Tree Protocol，多生成树协议）。

1. 二层网络转发环路问题分析

生成树协议的作用是为了解决二层交换网络的环路问题，那么什么情况下才会出现二层网络的环路问题呢？或者说环路是由于什么原因导致的？在部署交换网络时，为了提高网络的可靠性和冗余性，往往都会增加多台交换机并且在交换机之间部署多条链路级联。这样做的好处是如果其中一台交换机发生故障或者链路发生故障都不会影响网络的正常工作。这确实是一个非常好的增强冗余和可靠性的解决方案，但是这种做法同时也会带来转发环路问题。如图 14-1 所示的网络是一个交换环路网络，在这个网络中，三台交换机之间互连，在交换机 B（SWB）上连接了一台主机 A，在交换 C（SWC）上连接了一台主机 B。下面分析环路是怎样形成的。

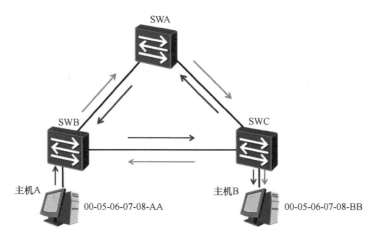

图 14-1 交换环路网络

首先，假设所有交换机的 MAC 地址表都为空，主机 A 和主机 B 通信，主机 A 将数据帧发送给交换机 B，交换机 B 收到数据帧后学习源 MAC 地址，并根据目的 MAC 地址查询 MAC 地址表进行转发，由于 MAC 地址表中没有目的 MAC 地址。交换机 B 会向除了接收端口以外的其他端口泛洪该数据帧。这样交换机 A 和交换机 C 就都能收到数据帧。它们会做和交换机 B 同样的操作，学习数据帧源 MAC 并泛洪数据帧，经过多次泛洪转发，数据帧又被发送回了交换机 B，交换机 B 收到后再继续泛洪，从而形成环路。我们可以思考一下，图 14-1 中的环路会一直持续下去吗？答案是不会的。因为这是一个由未知单播帧（如果单播数据帧的目的 MAC 地址不在 MAC 地址表中，那么该数据帧被称为"未知单播帧"）造成的环路，由于交换机 MAC 地址表没有这个单播数据帧的目的 MAC 地址，所以会一直泛洪数据帧。如果主机 B 收到数据帧后进行了回复，那么在交换机 C 上的 MAC 地址表中就存在主机 B 的 MAC 地址。交换机 C 在收到数据帧后会直接转发给主机 B 而不再进行泛洪处理，这样环路也就被消除了。但是如果同样的场景换成广播帧会怎么样？环路会一直持续下去，永不休止。二层数据帧不同于 IP 包，在 IP 包中可以通过 TTL 值防止三层 IP 包转发环路，而二层数据帧中没有这种机制，这就会造成二层转发的永久环路，极大地浪费了网络带宽和设备处理资源。

2．STP 的作用及工作原理

生成树协议的基本思想十分简单，在自然界中生长的树是不会出现环路的，如果网络也能够像一棵树一样生长就不会出现环路了。基于这种理念，在 STP 中定义了根桥（Root Bridge）、根端口（Root Port）、指定端口（Designated Port）、替代端口（Alternate Port）、路径开销（Path Cost）等概念，目的就在于以构建一棵树的方式来消除冗余二层网络中的转发环路，同时实现链路备份和路径最优化。用于构建这棵树的算法称为生成树算法（Spanning Tree Algorithm）。

要想实现这些功能，交换机之间必须要交互用于建立这棵树的信息，这些信息被称为 BPDU（Bridge Protocol Data Unit，桥协议数据单元）报文。BPDU 报文使用组播的方式进行传递，目的组播 MAC 地址为 01-80-C2-00-00-00。BPDU 报文中携带了用于计算生成树的所有信息，通过在交换机之间交互 BPDU 报文并利用生成树算法构建一棵二层网络的无环树。

STP 的工作流程是：首先选举根桥，根桥的角色就像树根一样，是一棵树最重要的部分。根桥通过比较桥 ID（Bridge ID）的大小进行选举，桥 ID 最小的交换机将成为交换网络中的根桥，在一个交换网络中只能有一个根桥，其他的交换机都为非根桥。接下来，所有非根桥交换机会各自选择一条"最粗壮"的树枝作为到根桥的最短路径，相应的连接端口被称为根端口。最后，在其他的"树枝"上（交换机间的互连链路）还需要进行"剪枝"操作，选举出一个指定端口。最终只有根端口和指定端口可以转发数据帧，而其他的端口全部被阻塞，不能转发数据帧。通过上述流程即可在交换网络中建立一棵无环树。

随着应用的深入和网络技术的发展，STP 的缺陷在应用中也被暴露了出来。STP 的缺陷主要表现在收敛速度上。

当拓扑发生变化后，整个网络需要重新执行生成树的收敛计算，而这个计算时延称为 Forward Delay，协议默认值是 15 秒。在所有交换机收到拓扑变化的消息之前，若旧拓扑结构中处于转发状态的端口还没有发现自己在新的拓扑中应该停止转发,则可能存在临时环路。为了解决临时环路的问题,STP 使用了一种定时器策略,即在端口从阻塞状态到转发状态中间加入侦听和学习状态,两次状态切换的时间长度都是 Forward Delay,这两种状态下交换机不转发任何数据帧,这样就可以保证在拓扑变化时不会产生临时环路。但是,这个看似良好的解决方案实际上带来的却是至少两倍 Forward Delay 的收敛时间,这在某些实时业务(如语音视频)中是不能被接受的。

3. STP 的选举

生成树协议通过在交换机之间交互 BPDU 报文来计算一棵无环树,图 14-2 中展示了 BDPU 报文格式。

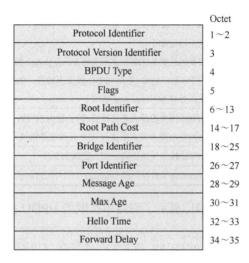

图 14-2 BPDU 报文格式

BPDU 报文格式说明参见表 14-1。

表 14-1 BPDU 报文格式说明

字段内容	说明
Protocol Identifier / 协议 ID	协议 ID="00"
Protocol Version Identifier / 协议版本 ID	协议版本标识符:STP 为 00;RSTP 为 02;MSTP 为 03
BPDU Type/BPDU 类型	STP BPDU 类型有以下两种: 0x00——STP 的配置 BPDU(Configuration BPDU) 0x80——STP 的拓扑改变通告 BPDU(TCN BPDU) RSTP/MSTP BPDU 类型: 0x02——RST BPDU(Rapid Spanning-Tree BPDU)或者 MST BPDU(Multiple Spanning-Tree BPDU)
Flags / 标志位	Flags 位由 8 比特组成,其中 STP 定义了其中 2 比特,分别是 Bit7 和 Bit0。Bit7 标识 TCA,用于确认接收到的 TCN BPDU 报文,Bit0 标识 TC,用于通告拓扑改变。STP 协议中只能在配置 BPDU 报文中设置 Flags 位

（续表）

字段内容	说明
Root Identifier / 根 ID	根 ID 是通告根桥的桥 ID，由 8 字节组成。前 2 字节由根桥优先级（优先级默认为 32768）+扩展系统 ID（VLAN ID）组成，后 6 字节标识根桥的背板 MAC 地址
Root Path Cost / 根路径开销	指 BPDU 报文的发送者到达根桥的距离，如果是根桥发送的 BPDU 报文，这个值为 0（根桥到自己的距离为 0）
Bridge Identifier / 桥 ID	BPDU 报文发送者的桥 ID，由 8 字节组成。前 2 字节由发送者的桥优先级（优先级默认为 32768）+扩展系统 ID（VLAN ID）组成，后 6 字节标识发送者的背板 MAC 地址
Port Identifier / 端口 ID	BPDU 报文发送者的端口 ID，由 2 字节组成：前 1 字节标识端口优先级，后 1 字节标识端口 ID。优先级默认为 128
Message Age / 消息老化时间	默认值为 20 秒，等于 Max Age 时间。BPDU 报文每经过一台交换机转发，该值会加 1，用于限制 BPDU 报文可以传递的范围。 判断过程如下： 交换机收到 BPDU 报文后会将 Message Age 和 Max Age 进行对比，如果 Message Age 值大于 Max Age，丢弃 BPDU 报文不转发；如果小于或等于 Max Age，转发 BPDU 报文，并将 Message Age 值加 1
Max Age / 最大老化时间	BPDU 报文的最大老化时间，默认值为 20 秒。如果超过 20 秒没有收到 BPDU 报文，认为网络出现故障重新执行 STP 计算收敛网络
Hello Time	BPDU 报文发送间隔时间，默认为 2 秒
Forward Delay / 转发延时	Listening（侦听）和 Learning（学习）两种状态的持续时间

图 14-3 所示为 BPDU 报文实例，展示了网络中抓取的 BPDU 报文格式及携带的数据内容。

图 14-3　BPDU 报文实例

STP 和 RSTP

IEEE 于 1998 年发布了 IEEE 802.1d 标准，定义了生成树协议——STP（Spanning Tree

Protocol），由于 STP 的收敛速度慢问题，2001 年发布了 IEEE 802.1w 标准，定义了快速生成树协议——RSTP（Rapid Spanning Tree Protocol），该协议是 STP 的优化协议，对原有的 STP 进行了更加细致的修改和补充。

STP 基本概念介绍如下。

（1）端口角色

STP 有 3 种端口角色：根端口（Root Port）、指定端口（Designated Port）、替代端口（Alternate Port）。

（2）端口状态

STP 有 5 种端口状态：禁用（Disabled）、侦听（Listening）、学习（Learning）、转发（Forwarding）和阻塞（Blocking）状态。

① Disabled：禁用状态，端口不处理和转发 BPDU 报文，也不转发数据帧。

② Listening：侦听状态，端口可以接收和转发 BPDU 报文，但不能转发数据帧。

③ Learning：学习状态，端口接收数据帧并构建 MAC 地址表，但不转发数据帧。增加 Learning 状态是为了防止未知单播数据帧造成临时环路。

④ Forwarding：转发状态，端口既可转发数据帧也可转发 BPDU 报文，只有根端口或指定端口才能进入 Forwarding 状态。

⑤ Blocking：阻塞状态，端口仅能接收并处理 BPDU 报文，但不转发 BPDU 报文，也不转发数据帧，此状态是替代端口的最终状态。

图 14-4 中描述了 STP 端口状态的转换。

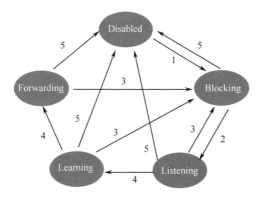

图 14-4　STP 端口状态的转换

STP 端口状态间的转换条件如下：

① 端口初始化或使能。

② 端口被选为根端口或指定端口。

③ 端口不再是根端口或指定端口。

④ Forward Delay 计时器超时（15 秒）。

⑤ 端口禁用或链路失效。

开启 STP 的交换机端口初始启动后，首先会从 Disabled 状态进入 Blocking 状态。在 Blocking 状态下，端口只能接收和分析 BPDU 报文，不能发送 BPDU 报文。如果端口被选举为根端口或指定端口,则会进入 Listening 状态,此时端口接收并发送 BPDU 报文,这种状态会持续一个 Forward Delay 的时间,默认值为 15 秒。然后,如果没有因"意外情况"而退回到 Blocking 状态,则该端口进入 Learning 状态,该状态同样会持续一个 Forward Delay 的时间。处于 Learning 状态的端口可以接收和发送 BPDU 报文,同时开始构建 MAC 地址表,为转发用户数据帧做好准备,但是处于 Learning 状态的端口仍然不能转发数据帧,只是构建 MAC 地址表。最后,端口由 Learning 状态进入 Forwarding 状态,开始转发数据帧。在整个状态的迁移过程中,端口一旦被关闭或者发生了链路故障,就会进入到 Disabled 状态。

(3)报文类型

STP 中定义了两种 BPDU 报文类型,分别是配置 BPDU(BPDU 报文中的 BPDU 类型值为 0x00)和 TCN BPDU(BPDU 报文中的 BPDU 类型值为 0x80),下面详解介绍这两种类型 BPDU 报文的区别。

1)配置 BPDU 报文

配置 BPDU 报文中 BPDU 类型(BPDU TYPE)的值为 0x00,主要作用如下所述。

① 用于选举根桥及端口角色。

② 通过定期(每隔 Hello Time 发送一次,默认为 2 秒)发送配置 BPDU 报文维护端口状态。

③ 用于确认接收到的 TCN BPDU 报文。

④ 用于选举根桥及端口角色。

开启 STP 的交换机初始化后,每台设备都会认为自己是根桥,分别发送配置 BPDU 报文选举根桥及端口角色。配置 BPDU 报文转发过程如图 14-5 中所示,STP 收敛后只有根桥才会定期发送配置 BPDU 报文,其他的非根桥接收后转发,通过这种方式维护端口状态。

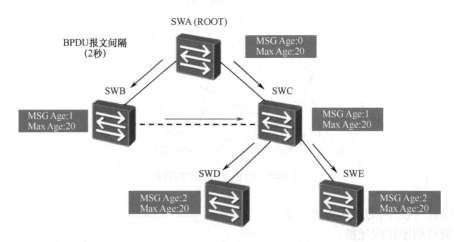

图 14-5 配置 BPDU 报文转发过程

由图 14-5 可知,STP 收敛后,SWA 被选举为根桥,每隔 Hello Time 发送一次配置 BPDU 报文,配置 BPDU 报文会从所有的指定端口发出,其他的非根桥从根端口接收到根桥发送的配置 BPDU 报文后,将配置 BPDU 报文缓存到接收端口,并将配置 BPDU 报文从所有的指定端口转发出去。但是非根桥在接收到配置 BPDU 报文后是否转发也会做出判断。

非根桥收到配置 BPDU 报文后，首先会将配置 BPDU 报文中的 Message Age 和 Max Age 做比对，如果 Message Age 小于或等于 Max Age，接收并转发配置 BPDU 报文。如果 Message Age 大于 Max Age 则会丢弃配置 BPDU 报文，不接收也不转发。而对于转发的配置 BPDU 报文会修改以下内容：

① 将配置 BPDU 报文中的桥 ID 修改为转发者的桥 ID。
② 将配置 BPDU 报文中端口 ID 修改为转发者的端口 ID。
③ 将配置 BPDU 报文中 Message Age 加 1（限制 BPDU 报文的传递范围）。

2）TCN BPDU 报文

TCN BPDU 报文中 BPDU 类型（BPDU TYPE）的值为 0x80，其作用是通告网络中的拓扑改变。首先需要说明通告 TCN BPDU 和 STP 的收敛没有任何关系，那么通告拓扑改变的目的是什么呢？我们先来看下面的场景，如图 14-6 所示的网络拓扑改变带来的问题（一）。

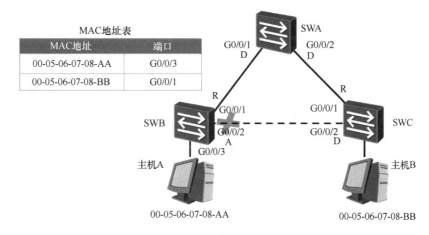

图 14-6　网络拓扑改变带来的问题（一）

在图 14-6 中，STP 收敛后 SWB 的 G0/0/2 端口被选举为替代端口（AP），端口被阻塞。主机 A 访问主机 B 的数据帧经过 SWB 转发给 SWA，在由 SWA 转发给 SWC。两台主机通信完后，SWB 的 MAC 地址表如图 14-6 所示。那么如果现在 SWA 和 SWC 之间的互连链路发生故障，会出现什么问题呢？继续看如图 14-7 所示的网络拓扑改变带来的问题（二）。

在图 14-7 中由于 SWA 和 SWC 之间的互连链路发生故障，导致 STP 重新收敛，收敛后的各端口角色如图 14-7 中所示。现在分析主机 A 访问主机 B 的数据帧是如何转发的，SWB 收到数据帧后通过查询 MAC 地址表将数据帧从 G0/0/1 端口转发出去，SWA 收到数据帧后会直接丢弃，丢弃的原因是链路故障造成端口关闭，数据帧无法再从 G0/0/2 端口发送出去，这样主机 A 和主机 B 也就无法通信了。那么主机 A 和主机 B 会一直无法通信吗？其实并不是，300 秒以后会发现主机 A 和主机 B 可以正常通信了。这是为什么呢？原因是等待 300 秒以后，SWB 上 G0/0/1 端口绑定主机 B 的 MAC 地址老化并被删除掉。删除后如果 SWB 又接收到访问主机 B 的数据帧，由于现在的 MAC 地址表中已经没有了主机 B 的 MAC 地址，这个数据帧将被从除接收端口以外的其他端口（G0/0/2）转发出去，这样 SWC 就收到了数据帧，主机 A 和主机 B 自然就恢复通信了。但是这种场景恢复时间太长了，每一次的拓扑变化都需要等待 300 秒后才能恢复通信，用户肯定是无法接受的。也许有人会说，在这种情况下可以通过将 MAC 地址表的老化时间改小解决问题。真的是这样吗？其实不然，这种解决方案根本就是治标不治本，并且会引发大量的未知单播帧泛洪。因为 MAC 地址表老化时间短，刚刚学习的 MAC 地址如果没有一个持续的访问流量，MAC 地址很快会被老化删除，再次收到同一单播帧后就会导致新一轮的泛洪，使网络不稳定。那有什么更好的方法能解决这个问题吗？答案是肯定的，就是通过 TCN BPDU 报文解决，网络拓扑改变带来的问题（三）如图 14-8 中所示，从中可以找到解决方案。

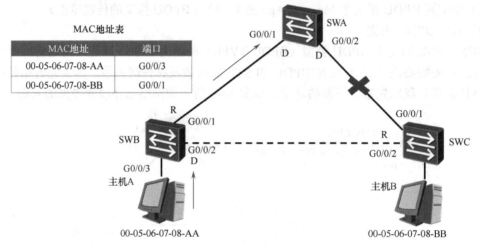

图 14-7 网络拓扑改变带来的问题（二）

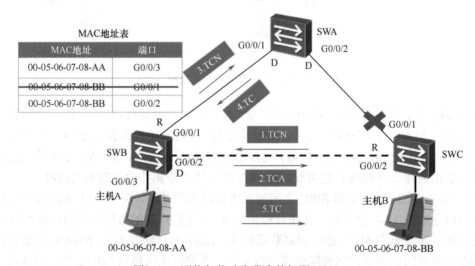

图 14-8 网络拓扑改变带来的问题（三）

① SWC 发现拓扑改变后，会从根端口发送一个 TCN BPDU 报文，目的是要将发生拓扑改变的消息通知根桥。

② SWB 从指定端口收到了 SWC 发送的 TCN BPDU 报文，SWB 会向 SWC 回复 TCA 置位的配置 BPDU 报文，用于确认接收到的 TCN BPDU 报文，如果 SWC 没收到上游设备回复的报文，则周期性地每隔 Hello Time 向根端口发送 TCN BPDU 报文。

③ SWB 继续从自己的根端口转发 TCN BPDU 报文给根桥。

④ SWA 收到 TCN BPDU 报文后同样向 SWB 回复一个 TCA 置位的配置 BPDU 报文，并将自己的 MAC 地址表老化时间修改为 15 秒，加速 MAC 地址老化。同时向所有的指定端口发送 TC 置位的配置 BPDU 报文，目的是告诉其他的非根桥拓扑已经发生了变化。该配置 BPDU 报文会连续发送 35 秒（Max Age + Forward Delay 的时间）。

⑤ 非根桥在收到 TC 置位的配置 BPDU 报文后会从所有的指定端口转发出去，同时将自己的 MAC 地址表老化时间修改为 15 秒，加速 MAC 地址老化。

注：在 STP 中 TC 置位的配置 BPDU 报文只能由根桥发送，而其他非根桥如果发现拓扑改变就需要以发送 TCN BPDU 报文的方式来告知根桥，再由根桥向全网发送 TC 置位的配置 BPDU 报文，目的是将所有交换机 MAC 地址表的老化时间修改为 15 秒，加速 MAC 地址老化，尽快恢复数据转发。

STP 协议发送 TCN BPDU 报文的条件：

① 端口从转发状态（Forwarding）过渡到阻塞状态（Blocking）或者禁用状态（Disabled）。
② 非根桥从一个指定端口收到 TCN BPDU 报文，会从自己的根端口进行转发。
③ 一个端口进入转发状态（Forwarding），并且本地已经存在一个指定端口。

（4）STP 收敛时间

STP 收敛完全依赖于定时器的计时，端口状态从 Blocking 状态迁移到 Forwarding 状态至少需要两倍的 Forward Delay 时间（需要 30 秒的时间），总收敛时间过长。STP 网络收敛后，如果直连链路发生故障，重新收敛需要 30 秒的时间，如果是次优配置 BPDU 报文或者非直连链路故障，则需要经过 50 秒的时间重新收敛。STP 直连链路故障收敛情况如图 14-9 中所示。

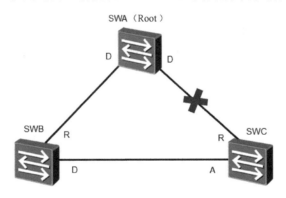

图 14-9　STP 直连链路故障收敛情况

由图 14-9 可知，如果 SWC 和 SWA 之间的互连链路发生故障，SWC 的替代端口会成为根端口，并且经过 30 秒的延时，端口过渡到转发状态。接下来再来看次优配置 BPDU 报文造成的 50 秒收敛延时，STP 非直连故障收敛情况如图 14-10 中所示。

由图 14-10 可知，如果 SWB 和 SWA 之间的互连链路发生故障，由于 SWB 连接根桥的端口关闭，SWB 会认为自己是根桥，并从指定端口发送配置 BPDU 报文，标识自己为根桥。SWC 从替代端口收到 SWB 发送的配置 BPDU 报文后，会和端口之前缓存的最优配置 BPDU 报文做对比，发现两个配置 BPDU 报文不一致，并且接收到的是一个次优配置 BPDU 报文，SWC 会直接忽略并继续等待接收端口缓存的配置 BPDU 报文。这样经过 20 秒的等待后超时，

SWC 端口角色重新收敛成为指定端口,并经过 30 秒的转发延时进入到转发状态,总收敛时间为 50 秒。

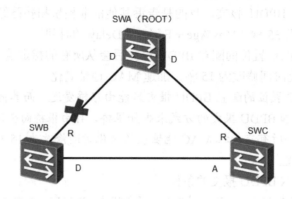

图 14-10 STP 非直连故障收敛情况

STP 的不足之处如下所述。

STP 虽然能够解决环路问题,但是由于网络拓扑收敛速度慢,影响了用户通信质量。如果网络中的拓扑结构频繁变化,网络也会随之频繁失去连通性,从而导致用户通信的频繁中断,使用户无法忍受。

14.1　在华为华三网络设备上配置标准生成树协议

图 14-11 所示为在园区网中实现标准 STP。如图 14-11 所示,在交换网络中实施生成树以防止二层环路。在交换网络中实施标准 STP,SW1 为业务 VLAN 的主根,SW2 为业务 VLAN 的备份根,由于在华为、华三设备上不支持基于每个 VLAN 的生成树,所以不能完成主、备根的实施。

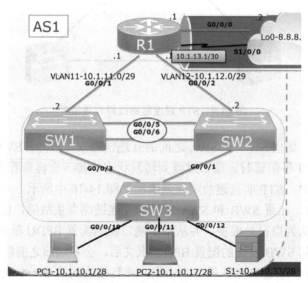

图 14-11　在园区网中实现标准 STP

同时在接入层交换机 SW3 上配置边缘端口功能和 BPDU 保护功能，以保护交换网络的生成树协议的稳定。

14.2 配置生成树协议要点

① 配置对应 VLAN、Trunk 和接入模式。
② 调整 STP 的模式，通过优先级调整 VLAN 的根设备。
③ 正确配置边缘端口。
④ 正确配置 BPDU 保护功能。

14.3 配置标准生成树协议详解

前置步骤：为了方便后续实验，请在交换机之间互连的链路上配置 Trunk，将 Trunk 上的 PVID 修改为 99，将连接 PC 的端口配置为 Access 模式，并依次接入业务 VLAN 8、VLAN 9、VLAN 10。

在 SW1、SW2、SW3 上创建 VLAN。
```
vlan batch 8 9 10 11 12 99
```
在 SW1、SW2、SW3 上的 Trunk 链路参考配置如下：
```
port link-type trunk
  port trunk pvid vlan 99
  port trunk allow-pass vlan 2 to 4094
```
SW3 的接入配置：
```
[SW3-GigabitEthernet0/0/10]interface GigabitEthernet0/0/10
[SW3-GigabitEthernet0/0/10] port link-type access
[SW3-GigabitEthernet0/0/10] port default vlan 8
[SW3-GigabitEthernet0/0/10]interface GigabitEthernet0/0/11
[SW3-GigabitEthernet0/0/11] port link-type access
[SW3-GigabitEthernet0/0/11] port default vlan 9
[SW3-GigabitEthernet0/0/11]int g0/0/12
[SW3-GigabitEthernet0/0/12] port link-type access
[SW3-GigabitEthernet0/0/12] port default vlan 10
```
验证 VLAN 端口的配置：
```
[SW3]display port vlan active
T=TAG U=UNTAG
```

Port	Link Type	PVID	VLAN List
GE0/0/1	trunk	99	U: 99
			T: 1 8 to 12
GE0/0/2	hybrid	1	U: 1
GE0/0/3	trunk	99	U: 99
			T: 1 8 to 12
GE0/0/4	hybrid	1	U: 1

Port	Type	PVID	VLAN
GE0/0/5	hybrid	1	U: 1
GE0/0/6	hybrid	1	U: 1
GE0/0/7	hybrid	1	U: 1
GE0/0/8	hybrid	1	U: 1
GE0/0/9	hybrid	1	U: 1
GE0/0/10	access	8	U: 8
GE0/0/11	access	9	U: 9
GE0/0/12	access	10	U: 10
GE0/0/13	hybrid	1	U: 1
GE0/0/14	hybrid	1	U: 1
GE0/0/15	hybrid	1	U: 1
GE0/0/16	hybrid	1	U: 1
GE0/0/17	hybrid	1	U: 1
GE0/0/18	hybrid	1	U: 1
GE0/0/19	hybrid	1	U: 1
GE0/0/20	hybrid	1	U: 1
GE0/0/21	hybrid	1	U: 1
GE0/0/22	hybrid	1	U: 1
GE0/0/23	hybrid	1	U: 1
GE0/0/24	hybrid	1	U: 1

14.3.1 修改生成树模式及生成树的根和备份根

修改生成树的模式为标准生成树，同时通过对优先级的修改使得汇聚层交换机 SW1 成为根设备。

在 SW1、SW2、SW3 上修改 STP 模式：

```
[SW2]stp mode ?
  mstp   Multiple Spanning Tree Protocol (MSTP) mode    //华为设备上默认为 MSTP, 多实例生成树
  rstp   Rapid Spanning Tree Protocol (RSTP) mode       //快速生成树
  stp    Spanning Tree Protocol (STP) mode              //标准生成树
[SW2]stp mode stp   //修改为标准的生成树，请读者自行在其他设备上修改。现网推荐使用 MSTP
```

查看 STP 的状态以判断设备的角色：

```
[SW1]dis stp brief
 MSTID  Port                Role  STP State      Protection
   0    GigabitEthernet0/0/1  DESI  FORWARDING     NONE
   0    GigabitEthernet0/0/3  DESI  FORWARDING     NONE
   0    GigabitEthernet0/0/5  ROOT  FORWARDING     NONE      //该端口为根端口，所以该端口连接的对端设备靠近根设备或者为根设备
   0    GigabitEthernet0/0/6  ALTE  DISCARDING     NONE
[SW2]display stp brief   //SW2 上所有端口显示为指定端口（DESI），状态为转发状态，所以判断 SW2 为该交换网络的根设备
 MSTID  Port                Role  STP State      Protection
   0    GigabitEthernet0/0/1  DESI  FORWARDING     NONE
   0    GigabitEthernet0/0/2  DESI  FORWARDING     NONE
   0    GigabitEthernet0/0/5  DESI  FORWARDING     NONE
   0    GigabitEthernet0/0/6  DESI  FORWARDING     NONE
```

```
[SW3]display stp brief
 MSTID   Port                    Role    STP State    Protection
   0     GigabitEthernet0/0/1    ROOT    FORWARDING   NONE     //存在根端口,所以
该端口连接的对端设备靠近根设备或者为根设备,这说明默认情况下 SW2 是根设备
   0     GigabitEthernet0/0/3    ALTE    DISCARDING   NONE
   0     GigabitEthernet0/0/10   DESI    FORWARDING   NONE
   0     GigabitEthernet0/0/11   DESI    FORWARDING   NONE
   0     GigabitEthernet0/0/12   DESI    FORWARDING   NONE
```

进一步根据各设备的 STP 详细情况,参考根交换机的选举规则进行判断。

```
[SW2]display stp
-------[CIST Global Info][Mode STP]-------
 CIST Bridge             :32768.4c1f-cca0-1ef8    //默认优先级为 32768,设备的 MAC 地址为
4c1f-cca0-1ef8
 Config Times            :Hello 2s MaxAge 20s FwDly 15s MaxHop 20
 Active Times            :Hello 2s MaxAge 20s FwDly 15s MaxHop 20
 CIST Root/ERPC          :32768.4c1f-cca0-1ef8 / 0
 CIST RegRoot/IRPC       :32768.4c1f-cca0-1ef8 / 0
 CIST RootPortId         :0.0
 BPDU-Protection         :Disabled
 TC or TCN received      :71
 TC count per hello      :0
 STP Converge Mode       :Normal
 Time since last TC      :0 days 0h:7m:13s
 Number of TC            :8
 Last TC occurred        :GigabitEthernet0/0/1
 ----[Port1(GigabitEthernet0/0/1)][FORWARDING]----
 Port Protocol           :Enabled
 Port Role               :Designated Port
 Port Priority           :128
 Port Cost(Dot1T )       :Config=auto / Active=20000
 Designated Bridge/Port  :32768.4c1f-cca0-1ef8 / 128.1
 Port Edged              :Config=default / Active=disabled
 Point-to-point          :Config=auto / Active=true
 Transit Limit           :147 packets/hello-time
 Protection Type         :None
[SW1]display stp
-------[CIST Global Info][Mode STP]-------
 CIST Bridge             :32768.4c1f-ccb7-3b11    //默认优先级为 32768,设备的 MAC 地址为
4c1f-ccb7-3b11
 Config Times            :Hello 2s MaxAge 20s FwDly 15s MaxHop 20
 Active Times            :Hello 2s MaxAge 20s FwDly 15s MaxHop 20
 CIST Root/ERPC          :32768.4c1f-cca0-1ef8 / 20000
 CIST RegRoot/IRPC       :32768.4c1f-ccb7-3b11 / 0
 CIST RootPortId         :128.5
 BPDU-Protection         :Disabled
 TC or TCN received      :37
 TC count per hello      :0
```

```
    STP Converge Mode           :Normal
    Time since last TC          :0 days 0h:8m:35s
    Number of TC                :8
    Last TC occurred            :GigabitEthernet0/0/5
    ----[Port1(GigabitEthernet0/0/1)][FORWARDING]----
     Port Protocol              :Enabled
     Port Role                  :Designated Port
     Port Priority              :128
     Port Cost(Dot1T )          :Config=auto / Active=20000
     Designated Bridge/Port     :32768.4c1f-ccb7-3b11 / 128.1
     Port Edged                 :Config=default / Active=disabled
     Point-to-point             :Config=auto / Active=true
     Transit Limit              :147 packets/hello-time
     Protection Type            :None
[SW3]display stp
-------[CIST Global Info][Mode STP]-------
 CIST Bridge                    :32768.4c1f-ccbd-35ba    //默认优先级为 32768，设备的 MAC 地址为
4c1f-ccbd-35ba
     Config Times               :Hello 2s MaxAge 20s FwDly 15s MaxHop 20
     Active Times               :Hello 2s MaxAge 20s FwDly 15s MaxHop 20
     CIST Root/ERPC             :32768.4c1f-cca0-1ef8 / 20000
     CIST RegRoot/IRPC          :32768.4c1f-ccbd-35ba / 0
     CIST RootPortId            :128.1
     BPDU-Protection            :Enabled
     TC or TCN received         :106
     TC count per hello         :0
     STP Converge Mode          :Normal
     Time since last TC         :0 days 0h:8m:15s
     Number of TC               :6
     Last TC occurred           :GigabitEthernet0/0/1
    ----[Port1(GigabitEthernet0/0/1)][FORWARDING]----
     Port Protocol              :Enabled
     Port Role                  :Root Port
     Port Priority              :128
     Port Cost(Dot1T )          :Config=auto / Active=20000
     Designated Bridge/Port     :32768.4c1f-cca0-1ef8 / 128.1
     Port Edged                 :Config=disabled / Active=disabled
     Point-to-point             :Config=auto / Active=true
     Transit Limit              :147 packets/hello-time
     Protection Type            :None
    ---- More ----
```

由于优先级相同，所以比较交换机的背板 MAC 地址，三台交换机的背板 MAC 地址分别为

- SW1：4c1f-ccb7-3b11；
- SW2：4c1f-cca0-1ef8；
- SW3：4c1f-ccbd-35ba。

很明显 SW2 的 MAC 地址更小，所以 SW2 为根交换机（注意在真机情况下，根设备会显示 this bridge is the root，而模拟器中不会显示）。

判断完毕根设备，做需求调整：

[SW1]stp priority 0　　//调整 SW1 的优先级为 0，0 是可用的优先级中最小的数值，SW1 最可能成为根设备

[SW1]dis stp brief　　//调整完毕后马上查看 STP 状态

MSTID	Port	Role	STP State	Protection
0	GigabitEthernet0/0/1	DESI	DISCARDING	NONE

//此时端口虽然为指定端口，但状态暂时为阻塞状态（即 Blocking 状态，DISCARDING 为华为设备上的表现形式），请回顾 STP 的 5 种状态

MSTID	Port	Role	STP State	Protection
0	GigabitEthernet0/0/3	DESI	DISCARDING	NONE
0	GigabitEthernet0/0/5	DESI	DISCARDING	NONE
0	GigabitEthernet0/0/6	DESI	DISCARDING	NONE

[SW2]stp priority 4096　　//SW2 作为备份根交换机，优先级调整为数值上第二小的数值

[SW2]dis stp brief

MSTID	Port	Role	STP State	Protection
0	GigabitEthernet0/0/1	DESI	DISCARDING	NONE
0	GigabitEthernet0/0/2	DESI	DISCARDING	NONE
0	GigabitEthernet0/0/5	ROOT	FORWARDING	NONE
0	GigabitEthernet0/0/6	ALTE	DISCARDING	NONE

STP 计算完毕后的状态：

[SW2]dis stp brief

MSTID	Port	Role	STP State	Protection
0	GigabitEthernet0/0/1	DESI	FORWARDING	NONE
0	GigabitEthernet0/0/2	DESI	FORWARDING	NONE
0	GigabitEthernet0/0/5	ROOT	FORWARDING	NONE
0	GigabitEthernet0/0/6	ALTE	DISCARDING	NONE

[SW3]dis stp brief

MSTID	Port	Role	STP State	Protection
0	GigabitEthernet0/0/1	ALTE	DISCARDING	NONE
0	GigabitEthernet0/0/3	ROOT	FORWARDING	NONE
0	GigabitEthernet0/0/10	DESI	FORWARDING	NONE
0	GigabitEthernet0/0/11	DESI	FORWARDING	NONE
0	GigabitEthernet0/0/12	DESI	FORWARDING	NONE

14.3.2　在接入层交换机上配置边缘端口

[SW3]stp edged-port default　　//配置交换设备所有端口为边缘端口，边缘端口使得 STP 跳过侦听、学习状态直接进入转发状态。边缘端口也可以在特定端口配置，后续会讨论

验证连接终端的端口：

[SW3]display stp interface g0/0/10
-------[CIST Global Info][Mode STP]-------
CIST Bridge　　　　　　　:32768.4c1f-ccbd-35ba
Config Times　　　　　　 :Hello 2s MaxAge 20s FwDly 15s MaxHop 20
Active Times　　　　　　 :Hello 2s MaxAge 20s FwDly 15s MaxHop 20
CIST Root/ERPC　　　　　 :0　　.4c1f-ccb7-3b11 / 20000

```
              CIST RegRoot/IRPC      :32768.4c1f-ccbd-35ba / 0
              CIST RootPortId        :128.3
              BPDU-Protection        :Disabled
              TC or TCN received     :143
              TC count per hello     :0
              STP Converge Mode      :Normal
              Time since last TC     :0 days 0h:15m:40s
              Number of TC           :8
              Last TC occurred       :GigabitEthernet0/0/3
              ----[Port10(GigabitEthernet0/0/10)][FORWARDING]----
               Port Protocol         :Enabled
               Port Role             :Designated Port
               Port Priority         :128
               Port Cost(Dot1T )     :Config=auto / Active=20000
               Designated Bridge/Port :32768.4c1f-ccbd-35ba / 128.10
               Port Edged            :Config=default / Active=enabled   //"enabled"表明该端口的边缘端口
```
特性配置已经使能，但是，是通过全局配置命令实现的，这一点体现在"default"上
```
               Point-to-point        :Config=auto / Active=true
               Transit Limit         :147 packets/hello-time
               Protection Type       :None
               Port STP Mode         :STP
               Port Protocol Type    :Config=auto / Active=dot1s
               BPDU Encapsulation    :Config=stp / Active=stp
               PortTimes             :Hello 2s MaxAge 20s FwDly 15s RemHop 20
               TC or TCN send        :0
               TC or TCN received    :0
               BPDU Sent             :3534
                       TCN: 0, Config: 3534, RST: 0, MST: 0
               BPDU Received         :0
                       TCN: 0, Config: 0, RST: 0, MST: 0
```
验证连接其他交换机的 Trunk 端口：
```
              [SW3]display stp interface g0/0/3
              -------[CIST Global Info][Mode STP]-------
              CIST Bridge            :32768.4c1f-ccbd-35ba
              Config Times           :Hello 2s MaxAge 20s FwDly 15s MaxHop 20
              Active Times           :Hello 2s MaxAge 20s FwDly 15s MaxHop 20
              CIST Root/ERPC         :0     .4c1f-ccb7-3b11 / 20000
              CIST RegRoot/IRPC      :32768.4c1f-ccbd-35ba / 0
              CIST RootPortId        :128.3
              BPDU-Protection        :Disabled
              TC or TCN received     :195
              TC count per hello     :0
              STP Converge Mode      :Normal
              Time since last TC     :0 days 0h:3m:25s
              Number of TC           :10
              Last TC occurred       :GigabitEthernet0/0/3
              ----[Port3(GigabitEthernet0/0/3)][FORWARDING]----
```

```
        Port Protocol              :Enabled
        Port Role                  :Root Port
        Port Priority              :128
        Port Cost(Dot1T )          :Config=auto / Active=20000
        Designated Bridge/Port     :0.4c1f-ccb7-3b11 / 128.3
        Port Edged                 :Config=default /   Active=enabled   //在华为设备上，连接交换机的 Trunk
端口也使能了边缘端口特性（注意 eNSP 存在一些 bug，如果关掉边缘端口再开启边缘端口，此时可能还会
显示没有开启边缘端口）
        Point-to-point             :Config=auto / Active=true
        Transit Limit              :147 packets/hello-time
        Protection Type            :None
        Port STP Mode              :STP
        Port Protocol Type         :Config=auto / Active=dot1s
        BPDU Encapsulation         :Config=stp / Active=stp
        PortTimes                  :Hello 2s MaxAge 20s FwDly 15s RemHop 0
        TC or TCN send             :1
        TC or TCN received         :18
        BPDU Sent                  :2
                 TCN: 1, Config: 1, RST: 0, MST: 0
        BPDU Received              :113
                 TCN: 0, Config: 113, RST: 0, MST: 0
```

验证边缘端口：

```
        [HW-SW3-GigabitEthernet0/0/10]shutdown    //关闭连接 PC1 的接入端口
        Dec 18 2018 17:28:54-08:00 HW-SW3 %%01PHY/1/PHY(l)[0]:         GigabitEthernet0/0/10: change
status to downun
        [HW-SW3-GigabitEthernet0/0/10]undo shutdown    //再次开启端口
        [HW-SW3-GigabitEthernet0/0/10]dis stp brief    //查看生成树的简要信息，会发现 GE0/01/10 端口
的状态马上成为转发状态，这就是边缘端口的功劳。本例没有进行数据测试，读者可以自行测试
        MSTID    Port                    Role      STP State      Protection
        0        GigabitEthernet0/0/1    ALTE      DISCARDING     NONE
        0        GigabitEthernet0/0/3    ROOT      FORWARDING     NONE
        0        GigabitEthernet0/0/10   DESI      FORWARDING     NONE
        0        GigabitEthernet0/0/11   DESI      FORWARDING     NONE
        0        GigabitEthernet0/0/12   DESI      FORWARDING     NONE
```

14.3.3　在接入层交换机上配置 BPDU 保护功能

请注意在华为设备上配置 BPDU 保护功能的先后顺序，请完成以下实验。

```
        [HW-SW3]stp bpdu-protection    //全局配置了 BPDU 保护功能，在华为设备上只能进行全局配置，
而不能在特定端口配置 BPDU 保护功能
        [HW-SW3]
        Nov 26 2018 20:30:31-08:00 HW-SW3 %%01MSTP/4/BPDU_PROTECTION(l)[0]:This edged-port
GigabitEthernet0/0/1 that enabled BPDU-Protection will be shutdown, because it received BPDU packet!
        Nov 26 2018 20:30:31-08:00 HW-SW3 %%01MSTP/4/BPDU_PROTECTION(l)[1]:This edged-port
GigabitEthernet0/0/3 that enabled BPDU-Protection will be shutdown, because it received BPDU packet!
        Nov 26 2018 20:30:32-08:00 HW-SW3 %%01PHY/1/PHY(l)[2]:         GigabitEthernet0/0/1: change
status to down
```

Nov 26 2018 20:30:32-08:00 HW-SW3 %%01PHY/1/PHY(l)[3]: GigabitEthernet0/0/3: change status to down

Nov 26 2018 20:30:32-08:00 HW-SW3 %%01IFNET/4/IF_STATE(l)[4]:Interface Vlanif1 has turned into DOWN state.

Nov 26 2018 20:30:32-08:00 HW-SW3 %%01IFNET/4/LINK_STATE(l)[5]:The line protocol IP on the interface Vlanif1 has entered the DOWN state

可以看到 SW3 上的很多端口都被关闭了，这是因为 SW3 开启了全局的边缘端口，导致所有端口都是边缘端口（包括 GE0/0/1 和 GE0/0/3 端口）。而此时 BPDU 保护功能已经开启，依据 BPDU 保护规则，这些端口被关闭。这当然会导致业务中断，那么解决方案是什么呢？请参照下面内容实施：

[HW-SW3-GigabitEthernet0/0/3]stp edged-port disable   //关闭边缘端口，非边缘端口就不在 BPDU 的保护范围中

[HW-SW3-GigabitEthernet0/0/3]shutdown

Info: Interface GigabitEthernet0/0/3 has already been shutdown.

[HW-SW3-GigabitEthernet0/0/3]undo shutdown   //被置于 err-disable 状态的端口，推荐先手动关闭端口再开启端口

Nov 26 2018 20:50:47-08:00 HW-SW3 %%01PHY/1/PHY(l)[3]: GigabitEthernet0/0/3: change status to up

[HW-SW3-GigabitEthernet0/0/1]stp edged-port disable

[HW-SW3-GigabitEthernet0/0/1]shutdown

Info: Interface GigabitEthernet0/0/1 has already been shutdown.

[HW-SW3-GigabitEthernet0/0/1]undo shutdown

所以，在配置华为的 STP 尤其是 BPDU 保护功能之前，应该先关掉特定端口（即交换机之间互连的 Trunk 端口）的边缘端口，再配置 BPDU 保护功能。

注意，此时 SW1 和 SW2 暂时还没有接入端口，但在后续架构中，两个核心交换机也会存在接入端口（即拓扑中 SW1 连接 R1 的端口以及 SW2 连接 R1 的端口），此时也需要配置边缘端口。至于 BPDU 保护功能，由于 SW1 和 SW2 是核心交换机，可以酌情暂不配置。推荐在端口下配置边缘端口。

SW1：

```
interface GigabitEthernet0/0/1
 port link-type access
 port default vlan 11
 stp edged-port enable
```

SW2：

```
<HW-SW2>dis cu int g0/0/2
#
interface GigabitEthernet0/0/2
 port link-type access
 port default vlan 12
 stp edged-port enable
[HW-SW2]display stp interface g0/0/2
-------[CIST Global Info][Mode STP]-------
CIST Bridge         :4096 .4c1f-cca0-1ef8
Config Times        :Hello 2s MaxAge 20s FwDly 15s MaxHop 20
Active Times        :Hello 2s MaxAge 20s FwDly 15s MaxHop 20
```

```
CIST Root/ERPC            :0      .4c1f-ccb7-3b11 / 10000
CIST RegRoot/IRPC         :4096 .4c1f-cca0-1ef8 / 0
CIST RootPortId           :128.1
BPDU-Protection           :Disabled
TC or TCN received        :19
TC count per hello        :0
STP Converge Mode         :Normal
Time since last TC        :0 days 0h:28m:9s
Number of TC              :4
Last TC occurred          :GigabitEthernet0/0/2
----[Port3(GigabitEthernet0/0/2)][FORWARDING]----
 Port Protocol            :Enabled
 Port Role                :Designated Port
 Port Priority            :128
 Port Cost(Dot1T )        :Config=auto / Active=20000
 Designated Bridge/Port   :4096.4c1f-cca0-1ef8 / 128.3
 Port Edged               :Config=enabled / Active=enabled    //Config=enabled 代表是端口配置使
                                                                能的边缘端口，这点有别于前边的
                                                                default 关键字
 Point-to-point           :Config=auto / Active=true
 Transit Limit            :147 packets/hello-time
 Protection Type          :None
 Port STP Mode            :STP
 Port Protocol Type       :Config=auto / Active=dot1s
 BPDU Encapsulation       :Config=stp / Active=stp
 PortTimes                :Hello 2s MaxAge 20s FwDly 15s RemHop 20
 TC or TCN send           :17
 TC or TCN received       :0
 BPDU Sent                :786
          TCN: 0, Config: 786, RST: 0, MST: 0
 BPDU Received            :0
          TCN: 0, Config: 0, RST: 0, MST: 0
```

通常情况下 STP 需要调整汇聚交换机成为根设备和备份根设备，在需要的位置配置边缘端口和 BPDU 保护功能。在华为设备上通用生成树不能实现每个 VLAN 实例，如果要实现基于特定 VLAN 的流量负载，需要多实例生成树 MSTP，这个知识点请参考笔者的 NP、IE 级别书籍。

14.3.4　华三设备生成树配置命令参考

在本节介绍一些华三设备的参考命令，这些命令的功能和前边描述的华为生成树功能一般相同，只是命令格式可能会稍有不同。

```
[H3C]stp enable
[H3C]stp mode ?
   mstp    Multiple spanning tree protocol mode
   rstp    Rapid spanning tree protocol mode
   stp     Spanning tree protocol mode
[H3C]stp mode stp //调整生成树模式
```

```
[H3C]display stp brief
    MSTID         Port              Role        STP State      Protection
     0          Ethernet1/0/1       DESI        LEARNING       NONE
[H3C]stp root primary     //设置主根设备，即调整优先级
[H3C]dis stp root
    MSTID    Root Bridge ID          ExtPathCost IntPathCost Root Port
     0      0.5cdd-70de-3d8d            0            0
[H3C]int e1/0/1
[H3C-Ethernet1/0/1]stp
[H3C-Ethernet1/0/1]stp ed
[H3C-Ethernet1/0/1]stp edged-port ?
    disable   Disable edge port
    enable    Enable edge port

[H3C-Ethernet1/0/1]stp edged-port en
[H3C-Ethernet1/0/1]stp edged-port enable    //配置边缘端口
Warning: Edge port should only be connected to terminal. It will cause temporary loops if port Ethernet1/0/1 is connected to bridges. Please use it carefully!

[H3C]display stp interface e1/0/1    //查看端口的 STP

----[CIST][Port1(Ethernet1/0/1)][FORWARDING]----
 Port Protocol              :enabled
 Port Role                  :CIST Designated Port
 Port Priority              :128
 Port Cost(Legacy)          :Config=auto / Active=200
 Desg. Bridge/Port          :0.5cdd-70de-3d8d / 128.1
 Port Edged                 :Config=enabled / Active=enabled
 Point-to-point             :Config=auto / Active=true
 Transmit Limit             :10 packets/hello-time
 Protection Type            :None
 MST BPDU Format            :Config=auto / Active=legacy
 Port Config-
  Digest-Snooping           :disabled
 Rapid transition           :false
 Num of Vlans Mapped        :2
 PortTimes                  :Hello 2s MaxAge 20s FwDly 15s MsgAge 0s RemHop 20
 BPDU Sent                  :100
          TCN: 0, Config: 100, RST: 0, MST: 0
 BPDU Received              :0
          TCN: 0, Config: 0, RST: 0, MST: 0
[H3C]stp bpdu-protection    //实施 BPDU 保护
[H3C]display stp brief
    MSTID         Port              Role        STP State      Protection
     0          Ethernet1/0/1       DESI        FORWARDING     BPDU
```

本案例实施完毕。

案例 15　思科的标准生成树协议

思科设备也默认开启生成树协议，不过不同于华为和华三设备默认开启 MSTP，思科设备默认开启的是 PVST（每个 VLAN 一棵生成树），每个 VLAN 一棵生成树拓扑意味着可以灵活地调整生成树的拓扑，而华为和华三设备要实现该功能需要使用 MSTP。生成树协议的三种模式和各厂商默认运行的模式对比如图 15-1 所示，白色字体代表默认的生成树模式。

	IEEE 802.1d	IEEE 802.1w	IEEE 802.1s
Cisco	PVST	Rapid PVST	MSTP
Huawei	STP	RSTP	MSTP
H3C	STP	RSTP	MSTP

图 15-1　生成树协议的三种模式和各厂商默认运行的模式对比

15.1　在思科设备上配置标准生成树协议

在思科设备上配置生成树协议拓扑如图 15-2 所示，在园区网的 SW1 到 SW3 上配置标准生成树协议以防止二层环路的产生。调整思科设备生成树模式，使得 SW1 成为 VLAN8、VLAN10、VLAN11 的根设备，SW1 成为 VLAN1、VLAN9、VLAN12 和 VLAN99 的备份根设备。

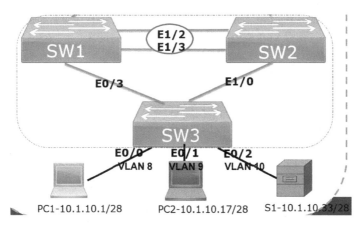

图 15-2　在思科设备上配置生成树协议拓扑

调整思科设备生成树，使得 SW2 成为 VLAN1、VLAN9、VLAN12 和 VLAN99 的根设备，SW2 成为 VLAN 8、VLAN10、VLAN11 的备份根设备，完成多个 VLAN 的互为主备架构。

在设备上配置边缘端口，从而加快生成树的收敛速度，提高网络品质。

在网络边界配置 BPDU 保护功能，保护生成树网络不受攻击。

15.2 案例拓扑说明

如图 15-2 所示，在 SW1、SW2 和 SW3 实现第 15.1 节的需求。其中 SW1 和 SW2 作为汇聚层交换机，会充当生成树的根设备和备份根设备，接入层交换机 SW3 也需要配置生成树协议，而且由于它更接近用户，所以在安全方面它更应该受到重视。

15.3 配置生成树协议要点

① 完成对应的交换机之间的 Trunk 和连接终端的接入模式设置。
② 首先确认 STP 模式和根设备，再进行其他配置。
③ 完成全局和端口的边缘端口配置。
④ 配置 BPDU 保护功能，查看因此产生的故障。

15.4 配置标准生成树协议步骤详解

15.4.1 调整生成树模式及根设备

请确保在完成本实验之前，所有交换机对应的 VLAN 已经创建，交换机之间的 Trunk 和连接终端的接入模式已经配置完成。

```
C-SW1(config)#spanning-tree mode pvst    //虽然思科设备默认是 PVST，但为了使读者加深理解，此处还是执行该命令，由于是默认命令，可能不会显示在 running-config。
C-SW1(config)#spanning-tree vlan 8,10,11 priority 0    //调整 VLAN8、VLAN10 和 VLAN11 的优先级为 0，使得 SW1 最可能成为根设备
C-SW1(config)#spanning-tree vlan 1,9,12,99 priority 4096    //优先级必须是 4096 的倍数，调整这 4 个 VLAN 的优先级为 4096，使得 SW1 成为这些 VLAN 的第二优先设备
C-SW2(config)#spanning-tree vlan 8,10,11 priority 4096    //对 SW2 做相应调整
C-SW2(config)#spanning-tree vlan 1,9,12,99 priority 0
```

验证生成树的状态，该命令可以让初学者快速认识生成树。

```
SW3#show spanning-tree vlan 8    //对于思科设备，运行的是 PVST，即每个 VLAN 都独立维护一棵生成树，所以可以确定地查验每个 VLAN 的 STP 状态
VLAN0008
  Spanning tree enabled protocol ieee
  Root ID    Priority    8    //SW3 得到的根交换机信息，优先级为 8，注意此处的"8"是由根设备的优先级 0 再加上扩展系统 ID（即 VLAN 号码）得到的
             Address     aabb.cc00.4000    //根交换机的背板 MAC 地址，即 SW1 的信息
             Cost        100    //到达根的开销为 100
             Port        4 (Ethernet0/3)    //E0/3 端口是根端口
             Hello Time  2 sec  Max Age 20 sec  Forward Delay 15 sec    //前文提及的 Hello Time（发送 BPDU 报文间隔）为 2 秒，最大生存时间为 20 秒，转发时延为 15 秒

  Bridge ID  Priority    32776   (priority 32768 sys-id-ext 8)    //SW3 自身的桥 ID 和优先级。可以看到优先级为 32776
```

```
                Address         aabb.cc00.6000   //SW3 自身的背板 MAC 地址，该地址不能改变
                Hello Time    2 sec   Max Age 20 sec   Forward Delay 15 sec
                Aging Time    300 sec

    Interface           Role  Sts  Cost       Prio.Nbr  Type
    ------------------- ----  ---  ---------  --------  --------------------------------
    E0/0                Desg  FWD  100        128.1     Shr
    E0/3                Root  FWD  100        128.4     Shr  //SW3 的 E0/3 口是根端口，处于转发状
态，到达根的开销为 100，端口的优先级为 128.4，端口类型为共享类型（因为该端口的速率为 10 Mbps）
    E1/0                Altn  BLK  100        128.5     Shr
```

在 SW3 上查验 VLAN9 的生成树信息：

```
    SW3#show spanning-tree vlan 9


    VLAN0009
      Spanning tree enabled protocol ieee
      Root ID    Priority    9    //在 SW3 上得到根的信息，根设备的优先级为 9，这个数值由优先
级 0+VLAN 号得出
                 Address     aabb.cc00.5000   //根设备的背板 MAC 地址，此处为 SW2 的 MAC 地址，
请读者自行查看
                 Cost        100
                 Port        5 (Ethernet1/0)   //E1/0 端口为根端口
                 Hello Time  2 sec   Max Age 20 sec   Forward Delay 15 sec

      Bridge ID  Priority    32777   (priority 32768 sys-id-ext 9)   //SW3 在 VLAN9 中的 STP 信息，
桥 ID 的优先级为 32768（默认）+VLAN 号 9
                 Address     aabb.cc00.6000
                 Hello Time  2 sec   Max Age 20 sec   Forward Delay 15 sec
                 Aging Time  300 sec

    Interface           Role  Sts  Cost       Prio.Nbr  Type
    ------------------- ----  ---  ---------  --------  --------------------------------
    E0/1                Desg  FWD  100        128.2     Shr
    E0/3                Altn  BLK  100        128.4     Shr  //在 VLAN9 中，E0/3 端口是处于阻塞状
态的非指定端口。请注意 E0/3 端口在 VLAN8 中是处于转发状态的，这说明在思科设备上，同一个端口针
对不同的 VLAN 可以处于不同的状态，这使得不同 VLAN 的数据可以在一定程度上实现负载分担
    E1/0                Root  FWD  100        128.5     Shr
```

生成树的根设备非常关键，必须保持其稳定性，否则会造成交换网络不稳定，这也是 STP 中必须调整的内容。

15.4.2 配置边缘端口

首先验证默认没有配置边缘端口时的故障情况。

```
    SW3(config)#int e0/0              //此处为连接 PC 的端口
    SW3(config-if)#shutdown           //关闭连接用户的端口，模拟网络出现故障后再恢复的场景
    SW3(config-if)#no shutdown        //马上开启端口
    SW3(config-if)#do sh spanning vlan 8   //验证 VLAN8 中生成树的状态
    Interface           Role Sts Cost       Prio.Nbr Type
```

```
                -------------------- ---- --- --------- -------- ----------------------------------
                E0/0                 Desg LIS 100           128.1         Shr   //此处仅给出了发生变化的端口，即 E0/0
端口，读者可以看到端口角色已经为指定端口（Desg），但是端口处于侦听状态（LIS），即此时数据依旧不
能转发
                R1-Cisco#ping 10.1.10.1    //由于生成树的原因，数据不能立刻转发。图 15-2 中没有显示 R1，依据
前边的整体拓扑，可以知道它的位置
                Type escape sequence to abort.
                Sending 5, 100-byte ICMP Echos to 10.1.10.1, timeout is 2 seconds:
                ......
                Success rate is 0 percent (0/5)
                R1-Cisco#ping 10.1.10.1    //经历 30 秒（即 2 个转发时延）之后，生成树状态切换为转发状态（FWD）
后，才转发用户的数据
                Type escape sequence to abort.
                Sending 5, 100-byte ICMP Echos to 10.1.10.1, timeout is 2 seconds:
                !!!!!
```

如果不配置边缘端口，会出现一些其他的故障，比如由于长时间无法转发报文导致 DHCP
无法获得有效地址等。

接下来在接入交换机上配置边缘端口。

```
                SW3(config)#spanning-tree portfast edge default   //在全局（非接口）配置边缘端口，使得所有接入
端口（不包含 Trunk 端口）成为边缘端口，这样可以快速地在多个接入端口完成配置工作。注意不同设备命
令可能稍有不同
                %Warning: this command enables portfast by default on all interfaces. You
                  should now disable portfast explicitly on switched ports leading to hubs,
                  switches and bridges as they may create temporary bridging loops.   //此处会有一个警告，边缘端口
特性在一个特殊环境中，比如 HUB 等连接的情况下可能会引发临时环路
                SW3(config)#
```

验证全局模式的边缘端口配置结果：

```
                SW3#show spanning-tree interface e0/0 detail
                 Port 1 (Ethernet0/0) of VLAN0008 is designated forwarding
                   Port path cost 100, Port priority 128, Port Identifier 128.1.
                   Designated root has priority 8, address aabb.cc00.4000
                   Designated bridge has priority 32776, address aabb.cc00.6000
                   Designated port id is 128.1, designated path cost 100
                   Timers: message age 0, forward delay 0, hold 0
                   Number of transitions to forwarding state: 1
                   The port is in the portfast edge mode by default
                   Link type is shared by default
                   BPDU: sent 406, received 0
```

思科设备不仅可以在全局配置边缘端口，还可以在特定端口配置边缘端口，而且选项中
也包含 Trunk，即针对 Trunk 也可配置边缘端口，该场景一般包括单臂路由（后面内容中会介
绍）中交换机的 Trunk 端口或者服务器虚拟化下的 Trunk 端口。

```
                C-SW1(config)#int e0/1
                C-SW1(config-if)#spanning-tree portfast ed
                C-SW1(config-if)#spanning-tree portfast edge ?
                   trunk   Enable portfast edge on the interface even in trunk mode
```

```
        <cr>
     C-SW1(config-if)#spanning-tree portfast edge    //在特定端口开启（关闭）边缘端口，同时可以对
Trunk 端口开启边缘端口
```

验证在端口下配置边缘端口后的 STP 状态：

```
    C-SW1#show spanning-tree interface e0/1 detail
     Port 2 (Ethernet0/1) of VLAN0008 is designated forwarding
       Port path cost 100, Port priority 128, Port Identifier 128.2.
       Designated root has priority 8, address aabb.cc00.4000
       Designated bridge has priority 8, address aabb.cc00.4000
       Designated port id is 128.2, designated path cost 0
       Timers: message age 0, forward delay 0, hold 0
       Number of transitions to forwarding state: 1
       The port is in the portfast edge mode   //该端口工作在边缘端口模式下，而且是端口配置，请对
比与前边介绍的全局配置的区别
       Link type is shared by default
       BPDU: sent 4670, received 0
```

请注意，边缘端口通常在接入交换机端口上全局配置，使得所有接入端口开启该特性。当在核心交换机上配置边缘端口时，推荐在某些需要的端口上配置，而不是全局配置。

15.4.3 配置 BPDU 保护功能

和华为设备不同，思科设备可以在全局和接入端口分别配置 BPDU 保护功能，其保护的范围有所不同。

全局配置的 BPDU 保护功能只有在开启边缘端口的端口上收到 BPDU 报文时才能起到保护作用，配置如下：

```
    SW3(config)#spanning-tree portfast edge default    //全局默认使能边缘端口
    SW3(config)#spanning-tree portfast edge bpduguard default   //全局 BPDU 保护对象针对的是边缘端口
```

为了方便演示故障的产生，接下来在端口配置 BPDU 保护功能。注意，端口下的 BPDU 保护只要求收到 BPDU 报文，而无论是否从边缘端口收到 BPDU 报文，都会将端口置于 err-disable 状态。

```
    SW3(config)#int e1/0
    SW3(config-if)#spanning-tree portfast edge trunk    //在端口上配置边缘端口,注意该端口是根端口或
者 Alternative 端口，注意不要在指定端口配置，因为指定端口在正常情况下只会发送 BPDU 报文
    SW3(config-if)#spanning-tree bpduguard enable    //对端口实施 BPDU 保护，该端口无论是否为边缘
端口都会生效
    *Nov 23 10:47:34.345: %SPANTREE-2-BLOCK_BPDUGUARD: Received BPDU on port E1/0 with
BPDU Guard enabled. Disabling port.    //该日志表明由于在 E1/0 端口上收到了 BPDU 报文，所以将端口关闭。
    SW3(config-if)#
    *Nov 23 10:47:34.345: %PM-4-ERR_DISABLE: bpduguard error detected on Et1/0, putting Et1/0 in
err-disable state
```

处于 err-disable 状态的端口需要手工关闭，然后再次开启或者自动恢复，如下所示：

```
    SW3(config-if)#no spanning-tree bpduguard enable
    SW3(config-if)#no shutdown    //直接开启端口无效
    SW3(config-if)#shutdown    //先手工关闭，再手动开启端口才会工作
```

```
SW3(config-if)#no shutdown
SW3(config-if)#
*Nov 26 12:23:12.050: %LINK-3-UPDOWN: Interface Ethernet1/0, changed state to up
*Nov 26 12:23:13.055: %LINEPROTO-5-UPDOWN: Line protocol on Interface Ethernet1/0, changed
state to up
```

请读者去掉该端口的 BPDU 保护功能，因为我们在一个不应该配置的位置配置了该功能。端口下的 BPDU 保护功能在存在多个需要配置的端口时比较烦琐，请读者根据情况设计和实施。

本案例实施完毕。

案例 16　华为设备以太网链路聚合

以太网链路聚合简书

在本书的描述中，以下称呼可以通用：以太网链路聚合、以太网链路捆绑、以太网聚合组。

将多个交换机以太网端口捆绑在一起所形成的组合称为以太网聚合组（又叫以太网链路聚合、以太网链路捆绑），而这些被捆绑在一起的以太网端口被称为该聚合组的成员端口。每个聚合组唯一对应着一个逻辑端口，我们称之为聚合端口。

以太网链路聚合示意图如图 16-1 所示。

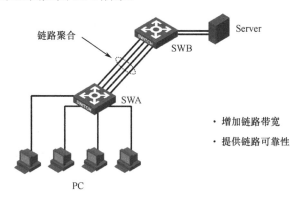

图 16-1　以太网链路聚合示意图

（1）聚合组／聚合端口类型

聚合组／聚合端口可以分为以下两种类型：

① 二层聚合组／二层聚合端口：二层聚合组的成员端口全部为二层以太网端口，其对应的聚合端口称为二层聚合端口（本案例仅演示二层聚合方式）。

② 三层聚合组／三层聚合端口：三层聚合组的成员端口全部为三层以太网端口，其对应的聚合端口称为三层聚合端口。在创建了三层聚合端口之后，还可以继续创建该三层聚合端口的子端口（简称三层聚合子端口）。三层聚合子端口也是一种逻辑端口，工作在网络层，主要用来在三层聚合端口上支持收发携带 VLAN TAG 的报文。

（2）LACP 简介

基于 IEEE 802.3ad 标准的 LACP（Link Aggregation Control Protocol，链路聚合控制协议）是一种实现链路动态聚合的协议，运行该协议的设备之间通过互发 LACPDU（Link Aggregation Control Protocol Data Unit，链路聚合控制协议数据单元）来交互链路聚合的相关信息。

（3）LACP 的基本功能

利用 LACPDU 的基本字段可以实现 LACP 的基本功能，基本字段包含以下信息：系统 LACP 优先级、系统 MAC 地址、端口聚合优先级、端口编号和操作 Key 等。

动态聚合组内的成员端口会自动使能 LACP，并通过发送 LACPDU 向对端通告本端的上述信息。当对端收到该 LACPDU 后，将其中的信息与本端其他成员端口收到的信息进行比较，选择能够处于选中状态的成员端口，使双方可以对各自端口的选中/非选中状态达成一致，从而决定哪些链路可以加入聚合组，以及某链路何时可以加入聚合组。可以理解为 $M:N$ 架构，M 即处于活动状态的链路，N 即处于备份状态，暂时没有加入以太网聚合的端口。

图 16-2 给出了 LACP 模式的以太网聚合活动和备份链路示意图。

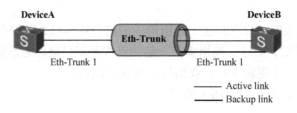

图 16-2　LACP 模式的以太网聚合活动和备份链路示意图

（4）LACP 的优先级

根据作用的不同，可以将 LACP 优先级分为系统 LACP 优先级和端口聚合优先级。

① 系统 LACP 优先级。

系统 LACP 优先级用于区分两端设备优先级的高低。要想使两端设备的选中端口一致，可以使一端具有较高的优先级，另一端则根据优先级较高的一端来选择本端的选中端口。优先级数值越小，优先级越高。

② 端口聚合优先级。

端口聚合优先级用于区分各成员端口成为选中端口的优先程度。

根据成员端口上是否启用了 LACP，可以将链路聚合分为静态聚合和动态聚合两种模式，介绍如下。

- 静态聚合模式：一旦配置好后，端口的选中/非选中状态就不会受网络环境的影响，比较稳定。但是不能根据对端的状态调整端口的选中/非选中状态，不够灵活。本例中将演示静态聚合方式（华为称之为手工模式，思科称之为"ON"，即手动模式）。
- 动态聚合模式：能够根据对端和本端的信息调整端口的选中/非选中状态，比较灵活，端口的选中/非选中状态容易受网络环境的影响，不够稳定。

处于静态聚合模式和动态聚合模式下的聚合组分别称为静态聚合组和动态聚合组。

通过改变负载分担的类型，可以灵活地实现聚合组流量的负载分担。用户既可以指定系统按照报文携带的 MAC 地址、服务端口号、报文入端口、IP 地址、MPLS 标签等信息之一或其组合来选择所采用的负载分担类型，也可以指定系统按照报文类型（如二层、IPv4、IPv6、MPLS 等）自动选择所采用的聚合负载分担类型，还可以指定系统对每个报文逐包进行聚合负载分担。通过采用不同的聚合负载分担类型及其组合，可以灵活地实现对聚合组内流量的负载分担。

除前面介绍的静态和动态聚合模式之外,思科还拥有自己私有的聚合协议,称为 PAgP(Port Aggregation Protocol)。

(5)实施以太网聚合的一般条件

- 成员端口必须是同一类型(相同的千兆位以太网口、相同的快速以太网口等)。
- 物理端口的数量、速率、双工方式、流控配置必须一致。
- VLAN 的端口模式相同,本征 VLAN 和允许的 VLAN 在 Trunk 上需要相同。
- 如果是接入端口,需要接入相同的 VLAN。
- 华为设备在实施端口聚合时必须为默认的 Hybrid 类型端口。
- 通常最多包含 8 个成员端口。
- 成员端口不能嵌套其他以太网聚合端口。

16.1 华为设备以太网链路聚合案例说明

以太网链路聚合可满足企业网内部流量日益增加的需求,充分利用现有设备和链路,即无须花费额外的费用来购买新的昂贵的硬件设备以及升级链路,同时它还可以消除生成树带来的链路浪费问题(在多链路互连时,生成树计算完毕后只有一条链路为转发状态)。另外,高速率端口也可以通过链路聚合方式进一步提高速率。

16.2 华为设备以太网链路聚合实施拓扑

图 16-3 所示为以太网链路聚合实施示意图,在图 16-3 中,将 SW1 与 SW2 互连的 G0/0/5 和 G0/0/6 捆绑为以太网链路聚合端口,从而使得 2 个汇聚层的核心交换机互连链路得到充分利用。

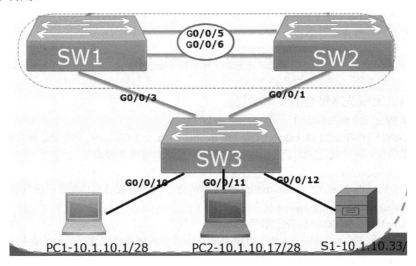

图 16-3 以太网链路聚合实施示意图

16.3 华为设备配置以太网链路聚合要点

① 华为设备需要在混杂端口下配置以太网聚合。
② 创建聚合端口（逻辑端口）。
③ 把物理端口加入逻辑聚合端口。
④ 查看以太网聚合端口的状态。

16.4 华为设备配置以太网链路聚合步骤详解

请注意本实验是在案例 14 的基础上进行的，即至少已经完成了 SW1、SW2、SW3 的 VLAN、接入模式和 Trunk 配置。

16.4.1 华为设备以太网链路聚合的前置条件和基本配置

在华为设备上配置以太网链路聚合相对比较麻烦，下面演示一个配置故障。

  [HW-SW2]int Eth-Trunk 12   //创建逻辑端口 Eth-Trunk，即以太网链路捆绑端口
  [HW-SW2-Eth-Trunk12]trunkport GigabitEthernet 0/0/5 to 0/0/6   //在逻辑端口中加入物理端口 G0/0/5 和 G0/0/6，即将 2 个物理端口捆绑为 1 个聚合端口
  Info: This operation may take a few seconds. Please wait for a moment...
  Error: The port has other configurations. Please clear them first.   //此处报错，是因为端口上存在其他配置，华为设备要求端口必须为混杂端口才可以配置以太网链路聚合，当然在配置以太网链路聚合后，可以再修改端口的类型
  The error port is GigabitEthernet0/0/5.   //该端口已经有配置存在，请清除配置
  [HW-SW2]clear configuration int g0/0/5
  Warning: All configurations of the interface will be cleared, and its state will be shutdown. Continue? [Y/N] :y   //清除端口 G0/0/5 的所有配置，恢复为默认配置，同时端口也会被关闭。注意并不是所有设备都支持该命令，只有在部分设备上可以执行该命令
  [HW-SW2]clear configuration int g0/0/6
  Warning: All configurations of the interface will be cleared, and its state will be shutdown. Continue? [Y/N] :y

华为网络以太网链路聚合的基本配置：

  [HW-SW2]int Eth-Trunk 12
  [HW-SW2-Eth-Trunk12]trunkport GigabitEthernet 0/0/5 to 0/0/6   //再次在聚合端口中加入 2 个物理端口，此次配置成功，该方式是华为设备上默认的手工配置以太网聚合方式
  Info: This operation may take a few seconds. Please wait for a moment...done.
  [HW-SW2-Eth-Trunk12]dis eth-trunk 12   //查看以太网链路聚合端口配置后的状态
  Eth-Trunk12's state information is:
  WorkingMode: NORMAL   //代表手工方式
  Hash arithmetic: According to SIP-XOR-DIP   //完成以太网链路聚合后，数据流中的源目 IP 地址被 hash 之后，根据 hash 值把数据转发到物理端口，这种负载均衡方式是默认方式，读者可以根据具体情况进行调整
  Least Active-linknumber: 1   Max Bandwidth-affected-linknumber: 8
  Operate status: down   //聚合端口此时并未工作

```
       Number Of Up Port In Trunk: 0    //聚合端口中没有工作的物理端口
       --------------------------------------------------------------
       PortName                    Status        Weight
       GigabitEthernet0/0/5        Down          1     //物理端口处于关闭状态，前边曾清空端口配置
导致端口被关闭
       GigabitEthernet0/0/6        Down          1
       [HW-SW2-GigabitEthernet0/0/5]int g0/0/5
       [HW-SW2-GigabitEthernet0/0/5]undo shut    //手工开启物理端口
       [HW-SW2-GigabitEthernet0/0/5]int g0/0/6
       [HW-SW2-GigabitEthernet0/0/6]
       [HW-SW2-GigabitEthernet0/0/6]undo sh      //手工开启物理端口
```

请读者自行完成 SW1 的配置，然后再次验证以太网链路聚合的配置状态。

```
       [HW-SW2]display eth-trunk 12
       Eth-Trunk12's state information is:
       WorkingMode: NORMAL            Hash arithmetic: According to SIP-XOR-DIP
       Least Active-linknumber: 1  Max Bandwidth-affected-linknumber: 8
       Operate status: up             Number Of Up Port In Trunk: 2   //有两个物理端口工作在以太网聚合
链路中
       --------------------------------------------------------------
       PortName                    Status        Weight
       GigabitEthernet0/0/5        Up            1
       GigabitEthernet0/0/6        Up            1
       [HW-SW2]display int Eth-Trunk 12    //查看以太网链路聚合端口
       Eth-Trunk12 current state : UP
       Line protocol current state : UP    //端口处于工作状态
       Description:
       Switch Port, PVID :     1, Hash arithmetic : According to SIP-XOR-DIP,Maximal BW: 2G, Current
BW: 2G, The Maximum Frame Length is 9216   //此时端口带宽累加到2G
       IP Sending Frames' Format is PKTFMT_ETHNT_2, Hardware address is 4c1f-cca0-1ef8
       Current system time: 2018-11-26 21:27:57-08:00
           Input bandwidth utilization  :    0%
           Output bandwidth utilization :    0%
       --------------------------------------------------------------
       PortName                    Status        Weight
       --------------------------------------------------------------
       GigabitEthernet0/0/5        UP            1
       GigabitEthernet0/0/6        UP            1
       --------------------------------------------------------------
       The Number of Ports in Trunk : 2
       The Number of UP Ports in Trunk : 2
```

16.4.2 以太网聚合端口在 STP 中的状态和 Trunk 模式配置

在完成以太网链路聚合之后，多个物理端口在逻辑上组成一个端口，由于该端口带宽增大，它在生成树中的开销会变小（更优化），当然这其中所有的物理端口都会处于转发状态。

[HW-SW2]dis stp brief　　//查看以太网聚合实施完成之后对 STP 拓扑的影响，此时 STP 中不再关注物理端口，而是把一个聚合端口作为生成树的端口来计算，如下所示，该端口是 STP 中的根端口，处于转发状态

```
MSTID  Port                      Role   STP State      Protection
  0    GigabitEthernet0/0/1      DESI   FORWARDING     NONE
  0    GigabitEthernet0/0/2      DESI   FORWARDING     NONE
  0    Eth-Trunk12               ROOT   FORWARDING     NONE
```

[HW-SW2]dis stp interface Eth-Trunk 12　　//查看以太网聚合逻辑端口在生成树中的状态
-------[CIST Global Info][Mode STP]-------
CIST Bridge   :4096 .4c1f-cca0-1ef8
Config Times   :Hello 2s MaxAge 20s FwDly 15s MaxHop 20
Active Times   :Hello 2s MaxAge 20s FwDly 15s MaxHop 20
CIST Root/ERPC   :0   .4c1f-ccb7-3b11 / 10000　　//请注意此处的开销值，华为设备默认的端口的开销值为 20000，而此时由于以太网链路聚合捆绑了 2 个物理链路，所以 STP 的开销值降为 10000
CIST RegRoot/IRPC   :4096 .4c1f-cca0-1ef8 / 0

注意，由于之前恢复了端口的默认配置，所以此时会影响用户的业务转发，需要把以太网聚合端口配置为 Trunk 模式，并允许对应的 VLAN 数据通过（所以在实施以太网聚合时，要非常小心，必须考虑到实施以太网聚合后是否会影响到用户的业务）。

下面的配置用于修改以太网聚合端口的 Trunk 模式和允许通过该端口的 VLAN 数据，这些配置会自动被物理端口继承（但不能在物理端口上实施）：

[HW-SW2]int Eth-Trunk 12
[HW-SW2-Eth-Trunk12]port link-type trunk
[HW-SW2-Eth-Trunk12]port trunk allow-pass vlan all
[HW-SW1]int Eth-Trunk 12
[HW-SW1-Eth-Trunk12]port link-type trunk
[HW-SW1-Eth-Trunk12]port tr allow-pass vlan all

华为设备以太网链路聚合实施完毕。

16.4.3　华三设备链路聚合命令参考

请读者参阅以下实施和验证命令，配置华三设备的静态方式以太网链路聚合，限于篇幅，以及为体现出本书对思科、华为设备配置的偏重性，后续会逐步减少甚至取消对华三设备配置的介绍。

[H3C]interface Bridge-Aggregation 1
[H3C-Bridge-Aggregation1]quit
[H3C]interface e1/0/1
[H3C-Ethernet1/0/1]port link-aggregation group 1
[H3C-Ethernet1/0/1]
[H3C-Ethernet1/0/1]quit
[H3C]display link-aggregation summary
Aggregation Interface Type:
BAGG -- Bridge-Aggregation, RAGG -- Route-Aggregation
Aggregation Mode: S -- Static, D -- Dynamic
Loadsharing Type: Shar -- Loadsharing, NonS -- Non-Loadsharing
Actor System ID: 0x8000, 5cdd-70de-3d8d

AGG Interface	AGG Mode	Partner ID	Select Ports	Unselect Ports	Share Type
BAGG1	S	none	1	0	NonS

本案例实施完毕。

案例 17　思科设备以太网链路聚合

在交换机互连端口实施手工模式的以太网链路聚合，增加交换机之间的可用带宽，同时增强可靠性（当部分链路失效时，其他端口可以继续工作）。思科设备以太网链路聚合实施拓扑如图 17-1 所示，在 SW1 和 SW2 上实施以太网链路聚合。

17.1　思科设备以太网链路聚合实施拓扑

思科设备以太网链路聚合实施拓扑如图 17-1 所示，在 SW1 与 SW2 之间的 E1/2 和 E1/3 端口上实施以太网链路聚合，将两个端口捆绑为一条逻辑链路，使得聚合链路拥有更宽的带宽，当然也会提高其冗余性，生成树也不会再阻塞以太网聚合链路中的物理链路，进一步提高链路的使用效率。

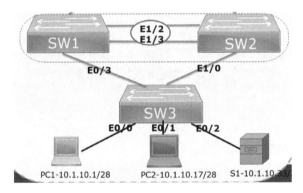

图 17-1　思科设备以太网链路聚合实施拓扑

17.2　思科设备以太网链路聚合配置要点

① 推荐在配置以太网链路聚合时，先关闭物理端口然后配置。
② 把物理端口加入对应以太网通道组（聚合端口）。
③ 开启物理端口并观察以太网链路聚合状态。

17.3　思科设备以太网链路聚合配置步骤详解

请注意本实验是在案例 15 的基础上进行的，即至少已经完成了 SW1、SW2、SW3 的 VLAN、接入模式和 Trunk 配置。

```
C-SW1(config)#int range e1/2 – 3    //进入连续的 E1/2 和 E1/3 端口
C-SW1(config-if-range)#shutdown    //建议在配置以太网链路聚合时关闭物理端口，如果不关闭端
```

口，在模拟器上可能会出现多种意想不到的 bug（在真机上偶尔也会如此）

```
C-SW1(config-if-range)#channel-group 12 mode on    //配置以太网聚合组 12，模式为手工模式
Creating a port-channel interface Port-channel 12
C-SW1(config-if-range)#no shutdown    //在另外一侧的设备上完成以太网链路聚合配置后，开启物理端口
!
C-SW2(config)#int r e1/2 - 3
C-SW2(config-if-range)#shutdown
C-SW2(config-if-range)#channel-group 12 mode on
Creating a port-channel interface Port-channel 12
C-SW2(config-if-range)#no shutdown
```

验证以太网链路聚合的状态：

```
C-SW2#show etherchannel summary
Flags:  D - down         P - bundled in port-channel
        I - stand-alone s - suspended
        H - Hot-standby (LACP only)
        R - Layer3       S - Layer2
        U - in use       N - not in use, no aggregation
        f - failed to allocate aggregator

        M - not in use, minimum links not met
        m - not in use, port not aggregated due to minimum links not met
        u - unsuitable for bundling
        w - waiting to be aggregated
        d - default port

        A - formed by Auto LAG
Number of channel-groups in use: 1
Number of aggregators:           1

Group  Port-channel  Protocol    Ports
------+-------------+-----------+-----------------------------------------------
12     Po12(SU)        -         Et1/2(P)    Et1/3(P)    //以太网聚合组12处于二层的可用状
```
态；"-" 代表没有使用任何动态协议，而是手工配置；2 个端口都被成功捆绑。如果是其他状态，请读者查阅上方的 Flags 去判断问题

```
C-SW2#show spanning-tree vlan 8   //查看 VLAN8 的 STP 配置内容


VLAN0008
  Spanning tree enabled protocol ieee
  Root ID    Priority    8
             Address     aabb.cc00.4000
             Cost        56
             Port        65 (Port-channel12)
             Hello Time  2 sec   Max Age 20 sec   Forward Delay 15 sec

  Bridge ID  Priority    4104   (priority 4096 sys-id-ext 8)
             Address     aabb.cc00.5000
```

```
                    Hello Time    2 sec   Max Age 20 sec   Forward Delay 15 sec
                    Aging Time    300 sec

Interface              Role Sts Cost          Prio.Nbr    Type
------------------- ---- --- --------- -------------------------------
Et1/0                  Desg FWD 100           128.5       Shr
Et1/1                  Desg FWD 100           128.6       Shr
Po12                   Root FWD 56            128.65      Shr      //两个物理端口捆绑成一个逻辑端口，
```
此时端口都处于转发状态，而在生成树中以 Po12，即 Port-channel 的逻辑端口出现，它是一个根端口，端口的 STP 开销为 56

```
C-SW1#show int trunk

Port       Mode        Encapsulation   Status      Native vlan
Et0/3      on          802.1q          trunking    99
Po12       on          802.1q          trunking    99      //原有物理端口的配置被继承到聚合链
```
路，思科设备不需要清空原有配置，只要原有配置不影响以太网链路聚合，就会自动继承

思科设备的以太网链路聚合配置完毕。

注意：如果在配置思科的链路聚合时出现 bug 而不能完成配置，可以尝试的解决方案如下。

① 删除已有的链路聚合；
② 关闭物理成员端口后重新配置链路聚合；
③ 修改链路聚合组的号码；
④ 如果有需要，可尝试重启设备。

至此，本案例实施完毕。

案例 18 交换机堆叠技术

交换机堆叠技术简书

各厂商都有自己的交换机堆叠技术（该技术最早起源于思科设备），这些技术大同小异，都可以把多个硬件交换机堆叠成一个在用户看来是一台交换机的技术。当然堆叠技术可以在核心汇聚交换机或者接入交换机上实施，它们拥有不同的名称。比如华为设备接入层的 iStack 技术，思科设备接入层的 FlexStack Plus、StackWise 技术等，再如在核心层交换机上实施框式设备的思科的 VSS（Virtual Switch System，虚拟交换机系统）技术，华为的 CSS（Cluster Switch System，集群交换系统）技术以及华三的 IRF（Intelligent Resilient Framework，智能弹性架构）技术等。

网络中主要存在两种形态的通信设备：盒式设备和框式设备。通常盒式设备部署在网络接入层或对可靠性要求不高的汇聚层，盒式单机设备对端口和带宽扩容不够灵活，扩容增加新的盒式设备会改变原组网结构，但它的优势也比较明显，就是投资成本相对较低。框式设备一般部署在网络核心层或汇聚层，具有高可靠性、高性能、高端口密度、可扩展性强的优点，由于投入成本较高，它不太适合部署在靠近用户侧的边缘网络。

华为设备上一种结合了两种设备优点的 iStack 堆叠技术应运而生。图 18-1 为 iStack 的堆叠示意图。iStack 堆叠就是将多台设备通过专用堆叠端口或业务端口连接起来，形成一台虚拟的逻辑设备，用户通过对这台虚拟设备进行管理来实现对堆叠中的所有设备的管理。华为设备支持两种模式的 iStack 堆叠：通过堆叠卡上专用堆叠端口进行堆叠的模式叫堆叠卡堆叠，它的主要优势是无须配置，直接连接专用堆叠端口就能实现 iStack 功能；通过业务端口堆叠的模式叫业务端口堆叠，它的主要优势是不需要专用堆叠卡，支持长距离堆叠。

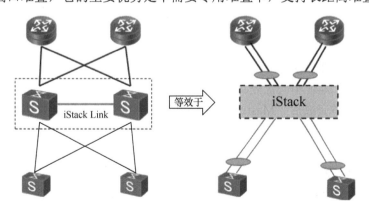

图 18-1 iStack 的堆叠示意图

（1）iStack 的角色

Master 设备：成员设备的一种，它负责管理整个堆叠。一个堆叠中同一时刻只能有一台成员设备成为 Master 设备。

① Standby 设备：成员设备的一种，Standby 设备是 Master 设备的备设备。当 Master 设备出现故障时，Standby 设备会接替原 Master 设备的所有业务。堆叠中只有一台 Standby 设备。

② Slave 设备：成员设备的一种，Slave 设备主要用于业务转发，它数量越多，堆叠系统的转发能力越强。堆叠中除了 Master 设备和 Standby 设备，其他设备都是 Slave 设备。

（2）物理连接

堆叠卡堆叠无须配置指定堆叠物理端口，堆叠卡上的端口是专用堆叠物理端口。

业务端口堆叠需要将业务端口配置成堆叠物理端口并加入堆叠端口，一台设备上只有两个堆叠端口，分别将其编号为 Stack-Port0 和 Stack-Port1。

堆叠物理端口之间可以使用专用堆叠线缆、光纤或标准网线连接，光纤可以将距离很远的物理设备连接成一台虚拟设备；标准网线连接以太网端口进行堆叠，百米以内可正常组网，使得组建堆叠更加灵活。

堆叠物理连接拓扑如图 18-2 所示。

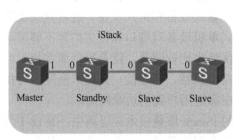
链形连接

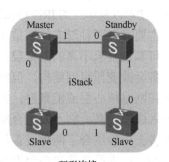
环形连接

图 18-2　堆叠物理连接拓扑

（3）角色选举

角色选举会在拓扑发生变化的情况下产生，比如，堆叠建立、新设备加入、堆叠分裂或两个堆叠合并。角色选举遵循以下原则：

- 系统运行时间长的优先；
- 成员优先级高的优先；
- 成员桥 MAC 地址小的优先。

Master 设备选举完成后，Master 设备会收集所有成员设备的拓扑信息，根据拓扑信息计算出堆叠转发表项和破坏点信息，下发给堆叠中的所有成员设备，并向所有成员设备分配堆叠 ID。

堆叠分裂是指在稳定运行的堆叠系统中带电移出部分成员设备，或者堆叠线缆多点故障导致一个堆叠系统变成多个堆叠系统。堆叠系统分裂之后需要做多主检测及冲突处理，保证业务继续稳定运行。

多主检测（Multi-Active Detection，MAD），是一种检测和处理堆叠分裂的协议，可降低堆叠分裂对业务的影响。

直连检测方式和代理检测方式：在同一堆叠系统中，两种检测方式互斥。直连检测方式指对堆叠成员设备间通过普通线缆直连的专用链路进行多主检测。在直连检测方式中，堆叠系统在正常运行时，不发送 MAD 报文；堆叠系统分裂后，分裂后的两台设备以 1 s 为周期通过检测链路发送 MAD 报文，从而进行多主冲突处理。

介绍完华为的 iStack 技术之后，我们来了解一下核心设备的硬件虚拟化。

VSS 是思科的一种网络系统虚拟化技术，将两台框式交换机（3850-48XS、4 系列、65 系列、76 系列、Nexus 系列）组合为单一虚拟交换机，从而提高运营效率、增强不间断通信能力。

将多台设备聚合成一台逻辑设备的技术，称为 CSS 技术。通过 CSS 将多台物理交换机聚合成一台逻辑交换机，提供更多槽位供业务使用。由于 CSS 将多台交换机聚合成单台交换机，原有上行的多条链路就变成了上行链路聚合组（Link Aggregation Group，LAG），不需要 MSTP 等环路协议，不需要阻塞冗余链路，提高了链路的利用率。简化网络架构的核心层堆叠技术如图 18-3 所示。

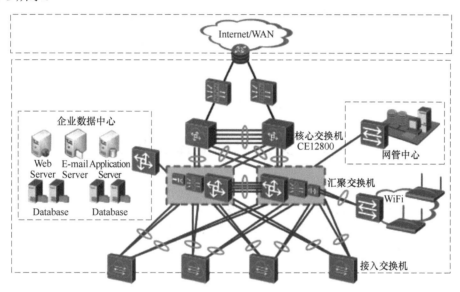

图 18-3　简化网络架构的核心层堆叠技术

在本案例中，将重点介绍华三的 IRF 技术，原因很简单，华三的 IRF 技术可以在其模拟器 HCL 上实施，而思科和华为的模拟器不支持堆叠技术。

IRF 的端口数目、交换容量是各台成员设备之和，IRF 技术极大地提高了系统的性能。IRF 堆叠中的成员设备分为 Master 和 Slave 设备，成员设备上用于堆叠连接的物理端口称为物理堆叠端口，物理堆叠端口需要和逻辑堆叠端口绑定，逻辑堆叠端口简称为堆叠端口，由多个物理堆叠端口聚合的堆叠端口称为聚合堆叠端口。

盒式交换机堆叠形成的虚拟设备等于一台框式分布式交换机，其中的 Master 设备作为主控板，Slave 设备作为备用主控板。

18.1　交换机堆叠案例实施说明

在核心层设备实施 IRF 组网，使得核心层性能和可靠性得到极大提高，另外可以熟悉 IRF

技术的实施命令,理解 IRF 实施成功后的多台交换机虚拟为一台交换机后的效果。

18.2 实施交换机堆叠拓扑

在设备 1 和设备 2 上实施 IRF 技术,如图 18-4 所示,图中两台设备通过 10 Gbps 端口互连。在本案例中使用华三的 HCL 模拟器(华为的 eNSP 模拟器和思科的 EVE 模拟器不支持堆叠命令)。

图 18-4　在设备 1 和设备 2 上实施 IRF 技术

18.3 交换机堆叠配置要点

① 规划好 IRF 角色(本例中设备 1 为 Master)。
② 关闭物理端口,把物理端口加入 IRF 逻辑端口。
③ 重置 Slave 设备的槽位,将物理端口加入 IRF 逻辑端口。
④ 在保存配置之后激活 IRF。

18.4 交换机堆叠配置步骤详解

18.4.1 配置 IRF 的 Master 设备优先级并加入物理端口

本部分命令在 IRF1 上执行,即在设备 1 上执行。

```
[IRF1]irf member 1 priority 32    //设置 IRF1(SW1)的 IRF 优先级为 32,该值已经是最大值,可
保证 IRF1 设备成为主设备(并未修改 IRF2 的优先级,意味着它使用默认的优先级 1)
[IRF1]int range te 1/0/50 to te 1/0/51    //进入连续的 10 Gbps 端口 1/0/50 和端口 1/0/51
[IRF1-if-range]shutdown    //关闭交换机互连端口,这是必需的步骤,请按规定操作
[IRF1-if-range]quit
[IRF1]irf-port 1/2    //设置 IRF1 的逻辑端口 1/2
[IRF1-irf-port1/2]port group interface te 1/0/50    //把物理端口 1/0/50 加入逻辑堆叠端口
You must perform the following tasks for a successful IRF setup:
Save the configuration after completing IRF configuration.
Execute the "irf-port-configuration active" command to activate the IRF ports.    //此处提示,必须按照
如下步骤来完成 IRF 配置,完成 IRF 配置后保存,然后执行 irf-port-configuration active 命令来激活 IRF
[IRF1-irf-port1/2]port g int te 1/0/51    //把物理端口加入逻辑端口
```

[IRF1-irf-port1/2]quit
[IRF1]int range te 1/0/50 to t 1/0/51
[IRF1-if-range]undo shdown　　//开启物理端口，注意此时并未激活 IRF，所以没有重启风险

18.4.2　配置 IRF 的 Slave 设备板卡序号

由于多个物理设备被虚拟化为一台设备，那么原先各个设备的板卡号需要按照新的序号编排，否则会出现重复的板卡号。

[IRF2]irf member 1 renumber 2　　//将 IRF2（即 SW2）的成员编号设置为 2，默认为 1
Renumbering the member ID may result in configuration change or loss. Continue?[Y/N]:y　　//对成员 1 的背板重新编号，该操作可能导致配置改变或者丢弃，是否继续？是
[IRF2]save　　//保存配置
The current configuration will be written to the device. Are you sure? [Y/N]:y　　//确定保存
Please input the file name(*.cfg)[flash:/startup.cfg]
(To leave the existing filename unchanged, press the enter key):　　//此处提示保存的文件名，使用默认的配置名称即可，华三设备即 start.cfg
Validating file. Please wait...
Saved the current configuration to mainboard device successfully.
[IRF2]quit
<IRF2>reboot　　//必须重启设备，重置的板卡接口序号才会生效
Start to check configuration with next startup configuration file, please wait.........DONE!
This command will reboot the device. Continue? [Y/N]:y　　//该命令将重启设备,确认设备重启完毕之后，继续配置 IRF
IRF2]int r t 2/0/50 to t 2/0/51　　//进入物理端口，注意之前已经"renumber"了该设备的板卡号码，所以此处接口为 10 Gbps 端口 2/0/50 和端口 2/0/51
[IRF2-if-range]shutdown　　//IRF2 设备重启后关闭物理端口
[IRF2-if-range]quit
[IRF2]irf-port 2/1　　//注意，推荐使用交叉互连，请对比 IRF1 设备，否则可能造成堆叠失败
[IRF2-irf-port2/1]port g int t 2/0/50　　//将物理端口 2/0/50 加入逻辑端口
You must perform the following tasks for a successful IRF setup:
Save the configuration after completing IRF configuration.
Execute the "irf-port-configuration active" command to activate the IRF ports.
[IRF2-irf-port2/1]port g int ten 2/0/51　　//将物理端口 2/0/51 加入逻辑端口
[IRF2-irf-port2/1]quit
[IRF2]int range ten 2/0/50 to t 2/0/51
[IRF2-if-range]undo shutdown　　//开启物理端口
[IRF2]save　　//保存配置
The current configuration will be written to the device. Are you sure? [Y/N]:y　　//确定保存配置
Please input the file name(*.cfg)[flash:/startup.cfg]
(To leave the existing filename unchanged, press the enter key):
flash:/startup.cfg exists, overwrite? [Y/N]:y　　//确定覆盖已经存在的配置，请键入 y

18.4.3　激活 IRF 的配置方法

在配置完毕前面的内容后，暂时还未激活 IRF，只有在激活之后，IRF 才会运行。当然在此过程中，设备会重启，之后虚拟化正式完成，如下所示。

<IRF1>save　　//保存配置，请在激活 IRF 前确定已经保存配置

```
[IRF1]irf-port-configuration active    //该命令用于激活 IRF 配置
[IRF2]%Nov 25 22:08:46:254 2018 IRF2 STM/6/STM_LINK_UP: IRF port 1 came up.    //由于 IRF1
已经激活 IRF，此时 IRF2 将会重启
<IRF2>system-view    //在 IRF 设备重启后，由于 RIF 没有配置完成，暂时不会立即显示 IRF1（即
IRF1 的主机名），此时依旧显示"IRF2"
System View: return to User View with Ctrl+Z.
[IRF2]int range ten 2/0/50 to t 2/0/51
[IRF2-if-range]shutdown    //手动关闭物理端口
[IRF2-if-range]und shu    //重新开启物理端口。此时设备可能再次重启
```

设备启动完毕，查看 IRF1 设备，如下所示。

```
[IRF1]display irf
MemberID    Role     Priority    CPU-Mac           Description
 *+1        Master   32          90c6-26d7-0104    ---    //IRF1 是 Master 设备，设备的 IRF 优先级
为 32，此处也显示了设备的背板 MAC 地址
  2         Standby  1           90c6-2ea5-0204    ---    //IRF2 也加入了 IRF1，它是 Standby 设备，
此时两个设备的资源共享，被其他设备看作一台设备
--------------------------------------------------
 * indicates the device is the master.
 + indicates the device through which the user logs in.

The bridge MAC of the IRF is: 90c6-26d7-0100
Auto upgrade              : yes
Mac persistent            : 6 min
Domain ID                 : 0
```

IRF 实施完毕，此时两台设备成为一台设备。

本案例实施完毕。

案例 19　华为华三设备的单臂路由技术

单臂路由技术简书

在前面的案例中讲到 VLAN 并不是为了完全隔离网络，而是为了使用户更好地通信，那么怎样能更好地管控通信呢？这就要采用 VLAN 间路由技术,使得处于不同 VLAN 中的设备可以通过路由的方式通信。

VLAN 在隔离广播的同时，也禁止了不同 VLAN 之间的用户通信，VLAN 路由（三层路由功能）成功解决了 VLAN 间用户设备的通信问题。

VLAN 间路由必须通过具备三层路由功能的设备完成，有以下几种解决方案。
- 路由器的单臂路由；
- 三层交换机 VLAN 接口 / 交换机虚拟接口；
- 不具备扩展功能的交换机的三层接口；
- 路由器的多臂路由（并不推荐）。

在本书中将会演示前三个解决方案，本例演示通过路由器来完成 VLAN 间路由的解决方案——单臂路由。

图 19-1 所示为单臂路由转发数据示意图。

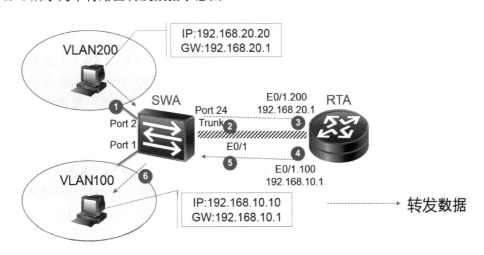

图 19-1　单臂路由转发数据示意图

划分 VLAN 后，VLAN 内的主机可以二层互通，而 VLAN 间的主机不能二层互通。可以在路由器上通过配置三层子接口的方式来实现 VLAN 间的三层互通，但如图 19-1 所示，当 Router 的三层以太网接口有限，只使用一个接口接入用户或网络时，一个接口上需要传输多个 VLAN 的报文，此时，可将一个三层以太网接口虚拟成多个逻辑子接口（相对于子接口而言，这个三层以太网接口称为主接口）。

单臂路由架构的通信过程如图 19-1 所示，位于 VLAN200 的终端 192.168.20.20 发送数据到 IP 地址为 192.168.10.10 的终端，首先需要把数据发送到网关 192.168.20.1。在❶处，二层交换机将在该数据帧上添加 PVID200 标签，该数据帧可以从❷处的 Trunk 直接携带 TAG200（即 VLAN200）发送，当该数据帧到达路由器时，路由器必须可以识别该数据帧，此时在❸处配置了路由器子接口（E0/1.200），基于 Dot1Q 200 的封装，路由器可以识别和接收该 VLAN 的流量。

在图 19-1 中，数据在到达路由器 RTA 后，路由器将根据流量的目的 IP 地址（192.168.10.1）通过直连路由的查找，把数据报文从❹处，即 E0/1.100 发出去，该子接口也完成了 Dot1Q 100 的封装，即该数据帧会携带 VLAN100 的 TAG；当该数据帧到达 SWA 处后，由于 SWA 的 E0/1 配置了 Trunk，所以可以识别该数据帧，对 TAG 不做任何改变；携带 TAG 100 的帧可以从接入接口（SW1 的 Port 1）剥除掉 TAG 后发送到 192.168.10.10，完成数据通信。

可以看到设备对接收到的报文中的 VLAN 标签进行识别，根据后续的转发行为对报文中的标签进行剥除，然后进行三层转发或其他处理。也就是为什么这些 VLAN 标签只在三层转发之前生效，之后的三层转发或其他处理不再依据报文中的这些标签进行。

VLAN 间通信的实质包含以下两方面：
- 对接口接收的报文，剥除 VLAN 标签后进行三层转发或其他处理；
- 对路由器接口发出的报文，又将相应的 VLAN 标签添加到报文中后再发送。

单臂路由的优点是可以节省接口数量和设备接口数量；缺点是链路复用，单点故障时易引发拥塞。通常适用于小型的分支网络或者办事处网络。

19.1 单臂路由实施案例及组网拓扑

华为设备上的单臂路由实施示意图如图 19-2 所示，在 AR3、SW4 上实施单臂路由，使得终端 Client1 和 Server 2 通过路由的方式实现通信，其中 SW4 作为接入层设备需要配置对应 VLAN 的接入模式和 Trunk 等，而 AR3 需要完成路由查找，最终实现不同 VLAN 间终端通信。

利用路由器 AR3 上一个接口（GE0/0/0）实现多个 VLAN（VLAN20 和 VLAN30）间主机的通信，可以相对节省路由器珍贵的接口资源。使得处于不同 VLAN 的 Client1 和 Server2 通过路由器的路由功能通信。当然两个终端的数据从 AR3 的 GE0/0/0 接口进入然后再次从 GE0/0/0 接口转发出去，这在大中型网络中并不实用，请根据具体网络选用。

19.2 单臂路由配置要点

单臂路由的配置要点如下：
① 配置路由器子接口，确保 VLAN 无误。
② 在交换机上创建 VLAN，完成连接主机的接入。

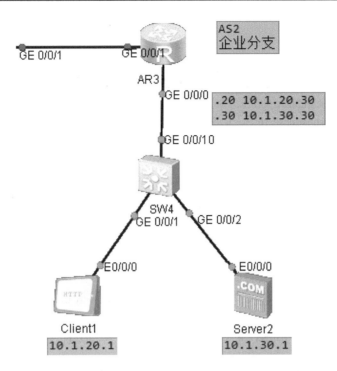

图 19-2 华为设备上的单臂路由实施示意图

③ 注意配置连接路由器的 Trunk 时允许特定 VLAN。
④ 给终端配置 IP 地址和网关进行数据测试。

19.3 配置单臂路由步骤详解

19.3.1 在路由器上配置子接口，标识特定 VLAN 的流量

在 AR3 设备上配置子接口，之后子接口将会自动创建直连路由，子接口最重要的功能是对 VLAN 数据的识别功能。

AR3 的配置如下：

```
    interface GigabitEthernet0/0/0.20    //配置标识为.20 的子接口,此处的数字可以和 VLAN 标识不同,但强烈推荐一致,以避免不必要的误解。
        dot1q termination vid 20    //标识 VLAN ID 的 TAG,此处表明 VLAN20 的流量
        ip address 10.1.20.30 255.255.255.224    //配置 VLAN20 的网关地址
        arp broadcast enable    //在模拟器中必须执行该命令,在大部分真机上也推荐执行该命令,否则路由器子接口不能主动发送 ARP 请求的广播报文,导致不同子网间的主机无法通信
    #
    interface GigabitEthernet0/0/0.30
        dot1q termination vid 30
        ip address 10.1.30.30 255.255.255.224
        arp broadcast enable
```

在配置完毕子接口后，请检查子接口是否工作，可以通过查看路由表来确认，重点查看

子接口产生的直连路由。

```
display ip routing
Route Flags: R - relay, D - download to fib
------------------------------------------------------------------------------
Routing Tables: Public
         Destinations : 10        Routes : 10

Destination/Mask       Proto   Pre   Cost    Flags   NextHop      Interface

      10.1.20.0/27     Direct  0     0       D       10.1.20.30   GigabitEthernet
0/0/0.20    //子接口产生了对应的直连路由
      10.1.20.30/32    Direct  0     0       D       127.0.0.1    GigabitEthernet
0/0/0.20
      10.1.20.31/32    Direct  0     0       D       127.0.0.1    GigabitEthernet
0/0/0.20
      10.1.30.0/27     Direct  0     0       D       10.1.30.30   GigabitEthernet
0/0/0.30    //子接口产生了对应的直连路由
      10.1.30.30/32    Direct  0     0       D       127.0.0.1    GigabitEthernet
0/0/0.30
      10.1.30.31/32    Direct  0     0       D       127.0.0.1    GigabitEthernet
0/0/0.30
      127.0.0.0/8      Direct  0     0       D       127.0.0.1    InLoopBack0
      127.0.0.1/32     Direct  0     0       D       127.0.0.1    InLoopBack0
127.255.255.255/32     Direct  0     0       D       127.0.0.1    InLoopBack0
255.255.255.255/32     Direct  0     0       D       127.0.0.1    InLoopBack
```

19.3.2 在交换机上配置 Trunk 和 Access 链路

为交换机连接路由器的接口配置 Trunk 接口，该 Trunk 链路用于承载多个 VLAN 的流量。

```
vlan batch 20 30    //创建 VLAN20 和 VLAN30
interface GigabitEthernet0/0/10
  port link-type trunk
  port trunk allow-pass vlan 2 to 4094    //该 Trunk 允许所有 VLAN 流量通过，如果需要精准控制可以仅允许 VLAN20 和 VLAN30 的流量通过
```

把连接用户的接口划分到对应的 VLAN20 和 VLAN30 中。

```
interface GigabitEthernet0/0/1
  port link-type access
  port default vlan 20
#
interface GigabitEthernet0/0/2
  port link-type access
  port default vlan 30
```

查看 Trunk 和 Access 接口的配置结果，以保证其正常工作。

```
[SW4]dis port vlan active
T=TAG U=UNTAG
```

Port	Link Type	PVID	VLAN List
GE0/0/1	access	20	U: 20
GE0/0/2	access	30	U: 30
GE0/0/3	hybrid	1	U: 1
GE0/0/4	hybrid	1	U: 1
GE0/0/5	hybrid	1	U: 1
GE0/0/6	hybrid	1	U: 1
GE0/0/7	hybrid	1	U: 1
GE0/0/8	hybrid	1	U: 1
GE0/0/9	hybrid	1	U: 1
GE0/0/10	trunk	1	U: 1 T: 20 30

19.3.3 配置 PC 的地址并完成数据测试

设置客户端 Client1 的 IP 地址掩码和网关，如图 19-3 所示。在本步骤中测试了网关的 IP 地址，这表明终端到网关的配置正确。

图 19-3 设置客户端 Client1 的 IP 地址掩码和网关

测试 Client1 到达另外一个终端网关的通信，如图 19-4 所示，结果表明到达另外一个终端的网络没有问题。

配置服务器的 IP 地址掩码及网关，如图 19-5 所示。

测试终端间通信，如图 19-6 所示。结果表明，客户端和服务器可以实现通信。

VLAN 间转发用户数据实验完成。该实验是一个小型的综合实验，它综合了 VLAN 技术、接入技术、Trunk 技术以及路由器的直连路由技术，所以几乎是入门的必备技术。

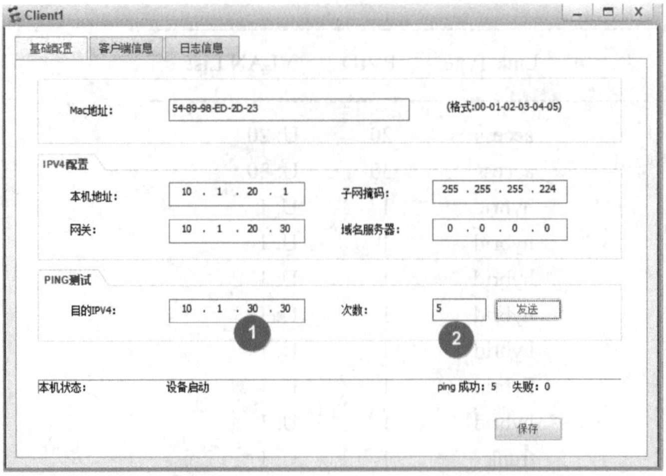

图 19-4　测试 Client1 到达另外一个终端网关的通信

图 19-5　配置服务器的 IP 地址掩码及网关

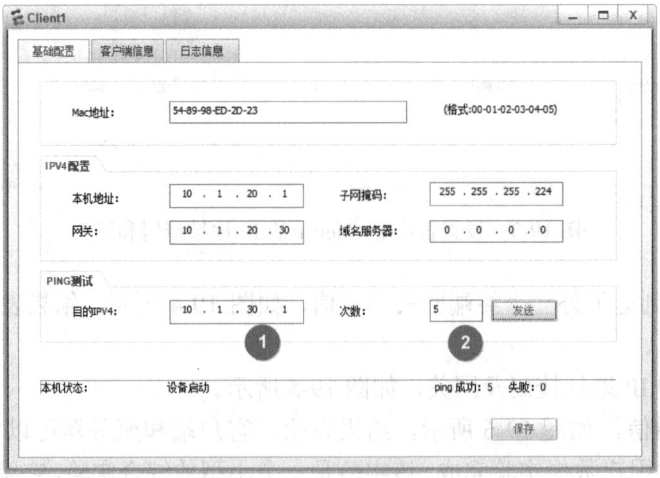

图 19-6　测试终端间通信

19.3.4 华三设备实现单臂路由命令参考

此处仅给出华三设备上对应的命令，分为交换机的配置命令和路由器的配置命令，请读者参考。

交换机的配置（创建 VLAN、配置接入接口和 Trunk 接口）：

```
vlan 20
port g1/0/1    //创建 VLAN20，并把端口 1 划分到 VLAN20
vlan 20
port g1/0/2
int g1/0/10
port link-type trunk
port trunk permit vlan all    //将上行端口设置成 Trunk 属性，允许 VLAN 透传
```

路由器的配置（开启物理接口，实施子接口）：

```
int e0/0
undo shutdown    //确保物理接口处于开启状态
int e0/0.20      //进入以太网的子接口 20
vlan-type dot1q vid 20    //在子接口里封装 VLAN10
ip address 10.1.20.30 27   //给子接口配置 IP 地址
int e0/0.30
vlan-type dot1q vid 30
ip address 10.1.30.30 27
```

案例 20 在思科设备上配置单臂路由

20.1 在思科设备上配置单臂路由案例

把转发决策表（即路由表）装载在路由器的子接口上，来完成多个 VLAN 间数据的访问，这就是单臂路由，关于其重要描述请参见案例 19。在本案例中，使用路由器的一个（物理）接口实现多个 VLAN（VLAN20 和 VLAN30）间主机的通信。

20.2 案例拓扑说明

单臂路由配置拓扑如图 20-1 所示，在 R3、SW4 上配置单臂路由，使得 PC3 和 S2 通信，在 R3 上创建两个子接口，交换机上连接的路由器使用干道模式（接口 0），在交换机连接终端的接口上实现对应的 VLAN 接入（E0/1 接口和 E0/2 接口）。

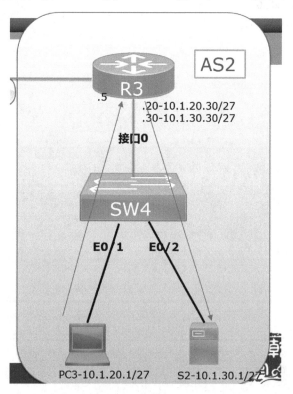

图 20-1 单臂路由配置拓扑

20.3 在思科设备上配置单臂路由步骤详解

20.3.1 在路由器上配置子接口，标识特定 VLAN 的流量

在本节主要配置子接口，请读者特别注意，思科路由设备上的接口默认为关闭状态，需要开启这些物理接口，否则子接口不会工作。

```
interface Ethernet0/0
  no ip address
  no shutdown            //思科设备的接口默认为关闭状态，需要开启逻辑子接口所在的物理接口，否则数据最终无法从子接口所属的物理接口发送
!
interface Ethernet0/0.20    //创建子接口.20，此处推荐接口 ID 和 VLAN ID 相同，避免冲突
  encapsulation dot1Q 20    //IEEE 802.1q 封装该子接口所在的 VLAN ID，此处为 VLAN20
  ip address 10.1.20.30 255.255.255.224    //在封装完毕后，配置子接口的 IP 地址，该地址是 VLAN20 中终端的网关地址
!
interface Ethernet0/0.30    //请读者自行配置.30 的子接口
  encapsulation dot1Q 30
  ip address 10.1.30.30 255.255.255.224
```

通过验证 IP 地址的简要信息和路由表查看配置结果，确保接口工作并且存在子接口产生的直连路由。

```
C-R3#show ip int brief
Interface            IP-Address      OK? Method Status                Protocol
Ethernet0/0          unassigned      YES NVRAM  up                    up
Ethernet0/0.20       10.1.20.30      YES manual up                    up
Ethernet0/0.30       10.1.30.30      YES manual up                    up
Ethernet0/1          unassigned      YES NVRAM  administratively down down
Ethernet0/2          unassigned      YES NVRAM  administratively down down
Ethernet0/3          unassigned      YES NVRAM  administratively down down
C-R3#show ip route
      10.0.0.0/8 is variably subnetted, 4 subnets, 2 masks
C        10.1.20.0/27 is directly connected, Ethernet0/0.20
L        10.1.20.30/32 is directly connected, Ethernet0/0.20
C        10.1.30.0/27 is directly connected, Ethernet0/0.30
L        10.1.30.30/32 is directly connected, Ethernet0/0.30
```

20.3.2 在思科交换机上配置 Trunk 和 Access 链路

在上行链路（连接路由器）的接口上配置 Trunk 链路，用于承载多个 VLAN 的流量。连接终端用户的接口完成对应 VLAN 的接入工作。

```
SW4(config)#vlan 20,30      //创建交换机转发数据所需的 VLAN20 和 VLAN30
SW4(config)#int e0/0
SW4(config-if)#switchport trunk encapsulation dot1q     //将连接路由器的接口配置为 Trunk 模式
SW4(config-if)#switchport mode trunk
```

```
SW4#show int trunk    //验证 Trunk 的配置结果

Port        Mode         Encapsulation    Status       Native vlan
Et0/0       on           802.1q           trunking     1           //Trunk 接口已经处于工作状
态，可以识别来自对端路由器子接口的携带 VLAN 标识的流量

Port        Vlans allowed on trunk
Et0/0       1-4094    //该 Trunk 默认允许所有 VLAN 的流量通过接口

Port        Vlans allowed and active in management domain
Et0/0       1,20,30

Port        Vlans in spanning tree forwarding state and not pruned
```

在连接终端（PC3 和 S2）的接口配置对应的接入 VLAN。

```
SW4(config)#int e0/1
SW4(config-if)#switchport mode access
SW4(config-if)#switchport access vlan 20
SW4(config-if)#int e0/2
SW4(config-if)#switchport mode access
SW4(config-if)#switchport access vlan 30
SW4#show vlan brief    //通过该验证命令，确保接口 E0/1 和 E0/2 已经划入 VLAN20 和 VLAN30
```

VLAN	Name	Status	Ports
1	default	active	Et0/3
20	VLAN0020	active	Et0/1
30	VLAN0030	active	Et0/2

20.3.3 完成终端配置和通信结果测试

在这一步骤中读者要特别注意，由于本实验中的 PC 是由路由器模拟的，所以需要关闭设备的路由功能，然后指定网关。

```
Server2 (config)#int e0/0
Server2 (config-if)#no sh    //开启物理接口
*Dec 25 06:18:41.065: %LINK-3-UPDOWN: Interface Ethernet0/0, changed state to up
*Dec 25 06:18:42.071: %LINEPROTO-5-UPDOWN: Line protocol on Interface Ethernet0/0, changed state to up
Server2 (config-if)#ip address 10.1.30.1 255.255.255.224
Server2(config)#no ip routing    //关闭路由器的路由功能
Server2(config)#ip default-gateway 10.1.30.30    //指定默认网关，这样去往其他网段的数据都将发送
到 10.1.20.30 的地址，然后由路由器进行路由决策并进行下一步转发
!
PC3(config)#int e0/0
PC3(config-if)#no sh
*Dec 25 06:19:22.710: %LINK-3-UPDOWN: Interface Ethernet0/0, changed state to up
*Dec 25 06:19:23.711: %LINEPROTO-5-UPDOWN: Line protocol on Interface Ethernet0/0, changed state to up
```

```
PC3(config-if)#ip address 10.1.20.1 255.255.255.224
PC3(config)#no ip routing
PC3(config)#ip default-gateway 10.1.20.30


PC3#show ip route
Default gateway is 10.1.20.30

Host               Gateway            Last Use    Total Uses   Interface
ICMP redirect cache is empty
Server2#show ip route   //查看对应的配置结果，会看到此时已经没有路由表，而只会显示默认网关
Default gateway is 10.1.30.30    //默认网关信息

Host               Gateway            Last Use    Total Uses   Interface
ICMP redirect cache is empty
```

在完成上述配置后继续进行测试：

```
Server2#ping 10.1.30.30    //测试终端到其所在网段网关的通信是否成功。此处已经完成通信
Type escape sequence to abort.
Sending 5, 100-byte ICMP Echos to 10.1.30.30, timeout is 2 seconds:
!!!!!
Server2#ping 10.1.20.30    //在终端和网关可以通信后，测试到其他网段所在网关的通信。此处完成
测试，表明路由器的路由表没有问题
Type escape sequence to abort.
Sending 5, 100-byte ICMP Echos to 10.1.20.30, timeout is 2 seconds:
!!!!!
Server2#ping 10.1.20.1    //最后测试终端与其他网段的终端通信。此处显示已经成功
Type escape sequence to abort.
Sending 5, 100-byte ICMP Echos to 10.1.20.1, timeout is 2 seconds:
!!!!!
```

至此，本案例实施完毕。

案例 21　华为华三网元的 VLANIF

VLANIF（SVI）技术简书

虽然前面案例中介绍了通过单臂路由实现 VLAN 间数据的通信，但是 VLANIF（华为、华三的称呼）或者 SVI（交换虚拟接口，思科的称呼）才是现实网络中使用最多的 VLAN 间互通技术。这些称呼在本书后续的描述中会通用。

请回忆前边的案例，在管理 VLAN 时，对一个 VLAN 赋予了 IP 地址，那么此时等同于在交换机上配置了一个地址，VLANIF 也是这样工作的，只要交换机支持配置 VLANIF 即可（部分二层交换机，仅支持配置一个 VLANIF）。通过 VLANIF 实现 VLAN 间三层互访有多种场景，常用的有同设备三层互访和跨设备三层互访。图 21-1 所示为同一交换机下的 VLAN 间数据互访，后面案例中会介绍跨设备的三层互访，把数据发送到更远的网络，两者的共同点是 VLANIF 的地址都是下面终端的网关或者同一网络的地址。

借助路由协议实现跨设备 VLAN 间数据通信如图 21-2 所示。

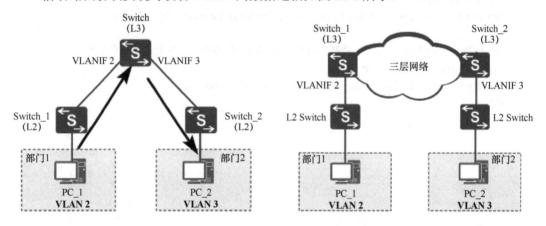

图 21-1　同一交换机下的 VLAN 间数据互访　　图 21-2　借助路由协议实现跨设备 VLAN 间数据通信

为了使得 VLANIF 工作，需要满足以下条件：① VLANIF 接口不能手动关闭；② 该 VLAN 中已经包含了物理接口（使用接入方式，Trunk 允许 VLAN 的数据通过，或者使用 HYBRID 方式允许 VLAN ID 通过），这些接口不能关闭。交换机的三层路由引擎支持 VLANIF/SVI，如图 21-3 所示，三层交换机内置的三层路由转发引擎执行路由功能。需要注意的是，华为和华三的三层交换机默认开启路由功能，而思科设备需要执行 ip routing 命令才可以出现 SVI 配置后的直连路由，以及其他路由表（模拟器默认开启）。

为什么在交换机上还要配置 IP 地址来划分这么多广播域呢？很简单，把管控范围控制在汇聚层交换机上，同时也只有这样才能去实施其他的技术和策略，比如 DHCP、VRRP、路由技术，等等。现网中很多交换机有了新的名字：路由交换机，路由器和交换机的功能被融合到一起，接口数也变得更多了。

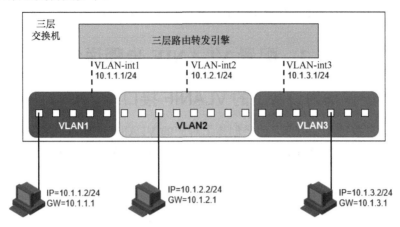

图 21-3 三层路由引擎支持 VLANIF/SVI

21.1 VLANIF 实施案例及拓扑

VLANIF 技术可以在多层交换机上创建针对某个 VLAN 的三层逻辑接口，这些接口作为该 VLAN 下终端的网关地址，它的灵活性使得该技术在企业园区网络中与单臂路由相比，处于统治地位。

图 21-4 为 VLANIF 在华为设备上的实施拓扑图，在 SW1 上配置 VLANIF 功能，使得位于 VLAN8 的 PC1 和位于 VLAN10 的 S1 可以实现通信。在 SW1 上配置 VLANIF8 和 VLANIF10，PC1 和 S1 的 IP 地址设置如图 2-4 中所示，VLAN8 接口和 VLAN10 接口的地址作为两个终端的网关设备地址。

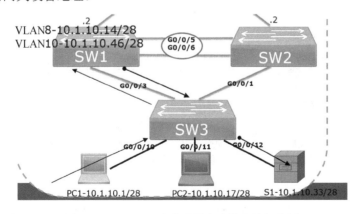

图 21-4 VLANIF 在华为设备上的实施拓扑图

21.2 VLANIF 配置要点

VLANIF 的配置要点如下。

① 创建 VLAN。
② VLAN 中存在接口，Trunk 允许 VLAN 流量通过或者配置了正确的接入模式。
③ 创建对应的 VLANIF，配置主机地址并配置正确的网关。

21.3 配置 VLANIF 步骤详解

21.3.1 配置华为设备上的 VLANIF 接口

请读者确认在执行此步骤前，已经创建了对应的 VLAN，完成了 Trunk 接口和 Access 接口的基本配置，读者可以参考案例 12。本例中仅给出 SW3 关于 VLAN 接口的验证。

在 SW3 上：

```
<SW3>display port vlan active
T=TAG U=UNTAG
--------------------------------------------------------------------
Port            Link Type       PVID    VLAN List
--------------------------------------------------------------------
GE0/0/1         trunk           1       U: 1
                                        T: 8 to 12 99    //交换机互连接口已经配置了 Trunk，并
且已经允许对应的 VLAN 流量通过该接口
GE0/0/2         hybrid          1       U: 1
GE0/0/3         trunk           1       U: 1
                                        T: 8 to 12 99
GE0/0/4         hybrid          1       U: 1
GE0/0/5         hybrid          1       U: 1
GE0/0/6         hybrid          1       U: 1
GE0/0/7         hybrid          1       U: 1
GE0/0/8         hybrid          1       U: 1
GE0/0/9         hybrid          1       U: 1
GE0/0/10        access          8       U: 8
GE0/0/11        access          9       U: 9
GE0/0/12        access          10      U: 10       //连接主机的接口已经划分到对应的 VLAN 中，
该接口为 VLAN10 接口
```

在 SW1 上配置 VLANIF 接口：

```
interface Vlanif8
 ip address 10.1.10.14 255.255.255.240    //创建 VLANIF8 接口，配置地址，注意该地址是所处网段
的最大可用地址，看到这里可以复习一下 IP 地址子网的配置
#
interface Vlanif10
 ip address 10.1.10.46 255.255.255.240    //在 SW1 上创建对应的 VLAN10，华为设备上的 VLANIF 默
认处于开启状态
```

VLAN 接口的工作状态和以下因素有关：VLAN（是否存在且工作）、Trunk（是否允许对应 VLAN 的流量通过）、ACCESS（是否把接口划分到了对应的 VLAN 中）、VLANIF 开启接口、VLANIF 接口是否工作（display ip int brief），以及其对应的直连路由是否存在，这一点非常关键，如下查看对应的直连路由已经存在于路由表中。

```
[HW-SW1]display ip routing-table
Route Flags: R - relay, D - download to fib
------------------------------------------------------------------------------
Routing Tables: Public
         Destinations : 6        Routes : 6

Destination/Mask    Proto    Pre  Cost    Flags    NextHop        Interface

    10.1.10.0/28    Direct   0    0       D        10.1.10.14     Vlanif8
    10.1.10.14/32   Direct   0    0       D        127.0.0.1      Vlanif8
    10.1.10.32/28   Direct   0    0       D        10.1.10.46     Vlanif10
    10.1.10.46/32   Direct   0    0       D        127.0.0.1      Vlanif10
```

21.3.2 华为设备数据测试和华三命令参考

数据测试过程如下所示，推荐读者测试的步骤为先测试本网络的网关通信，再测试到其他网络的网关通信，最后测试到其他终端的通信。

```
PC1>ping 10.1.10.14

ping 10.1.10.14: 32 data bytes, Press Ctrl_C to break
From 10.1.10.14: bytes=32 seq=1 ttl=255 time=78 ms
From 10.1.10.14: bytes=32 seq=2 ttl=255 time=63 ms

--- 10.1.10.14 ping statistics ---
   2 packet(s) transmitted
   2 packet(s) received
   0.00% packet loss
   round-trip min/avg/max = 63/70/78 ms

PC1>ping 10.1.10.46

ping 10.1.10.46: 32 data bytes, Press Ctrl_C to break
From 10.1.10.46: bytes=32 seq=1 ttl=255 time=78 ms
From 10.1.10.46: bytes=32 seq=2 ttl=255 time=47 ms

--- 10.1.10.46 ping statistics ---
   2 packet(s) transmitted
   2 packet(s) received
   0.00% packet loss
   round-trip min/avg/max = 47/62/78 ms

PC1>ping 10.1.10.33     //测试表明，不同网络间的终端通过采用三层交换机的路由决策完成了通信

ping 10.1.10.33: 32 data bytes, Press Ctrl_C to break
From 10.1.10.33: bytes=32 seq=1 ttl=254 time=157 ms
From 10.1.10.33: bytes=32 seq=2 ttl=254 time=109 ms
```

```
--- 10.1.10.33 ping statistics ---
2 packet(s) transmitted
2 packet(s) received
0.00% packet loss
round-trip min/avg/max = 109/133/157 ms
```

配置完 VLANIF 接口后，就可以实现终端到网关的通信，以及终端到其他 VLAN 中的终端通信了。那么如何实现远程网络的通信呢？现在使用三层交换机或者路由器的直连路由进行的数据转发决策，后续会使用静态路由或者采用动态路由协议得到其他形式的路由，然后再进行数据转发。

只有在完成本实验之后，才可以完成后续的路由技术、VRRP 技术等的学习，所以请熟练掌握本部分内容。读者需要自行完成后续案例中多个交换机的 SVI 配置。

在本部分华三的实施命令和华为几乎相同，命令如下所示，不再赘述。

```
[Switch] interface vlan-interface 1
[Switch-Vlan-interface1] ip address 172.16.1.1 255.255.255.0
```

本案例实施完毕。

案例 22　思科网元的 SVI 技术

22.1　使用 SVI 技术组网案例及实施拓扑

思科实施 SVI 拓扑如图 22-1 所示，通过在 SW1 和 SW2 上配置交换机虚拟接口（SVI），使得位于 VLAN8 的 PC1 和位于 VLAN10 的 S1 可以实现通信。PC1 和 S1 的地址如图 22-1 所示，10.1.10.14 是 PC1 的网关，10.1.10.46 是 S1 的网关。

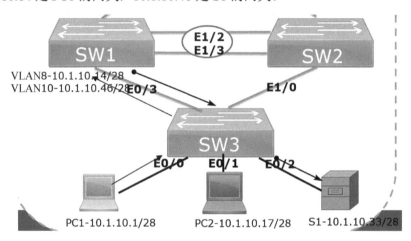

图 22-1　思科实施 SVI 拓扑

22.2　SVI 技术配置要点

SVI 技术配置要点如下。
① 创建 VLAN。
② VLAN 中存在接口，或者 Trunk 允许 VLAN 的流量通过或者配置了正确的接入模式。
③ 创建对应的 SVI，注意，在思科设备中需要手动开启 SVI。
④ 配置主机 IP 地址，关闭路由功能，并配置正确的网关。

22.3　配置 SVI 步骤详解

22.3.1　配置 SVI 的前置条件

创建对应 VLAN，把对应接口进行 VLAN 划分并配置交换机之间的 Trunk，该部分操作请参照案例 11。

在 SW3 上完成相关配置工作：

```
interface Ethernet0/0
  switchport access vlan 8
  switchport mode access
!
interface Ethernet0/1
  switchport access vlan 9     //此处也配置了关于 VLAN9 的接入接口，这是为了后续实验做准备，读者在本部分实验中可以忽略
  switchport mode access
!
interface Ethernet0/2
  switchport access vlan 10
  switchport mode access
SW3#show vlan brief     //对应 VLAN8 和 VLAN10 的接口已经划分好，在后续实验中，我们将不再讨论基本的 VLAN 配置，请读者自行配置
```

VLAN	Name	Status	Ports
1	default	active	Et1/1, Et1/2, Et1/3
8	QCNA	active	Et0/0
9	VLAN0009	active	Et0/1
10	VLAN0010	active	Et0/2

```
SW3#show interfaces trunk
Port       Mode          Encapsulation    Status        Native vlan
Et0/2      on            802.1q           trunking      1
Et0/3      on            802.1q           trunking      1

Port       Vlans allowed on trunk
Et0/2      1-4094
Et0/3      1-4094    //Trunk 上默认允许所有 VLAN 的流量通过

Port       Vlans allowed and active in management domain
Et0/2      1,8-12,99
Et0/3      1,8-12,99

Port       Vlans in spanning tree forwarding state and not pruned
Et0/2      1,8-12,99
Et0/3      1,8-12,99
```

22.3.2　在交换机上配置 SVI

在实施本步骤之前，请再次确认之前 VLAN 的相关接口接入或 Trunk 接口可正常工作。

```
SW1(config)#ip routing    //真机默认不开启三层交换机的路由功能，需要手动开启路由功能。模拟器仅供大家练习使用，模拟器默认开启路由功能
SW1(config)#int vlan 8
SW1(config-if)#ip address 10.1.10.14 255.255.255.240
SW1(config-if)#no shutdown   //在思科设备上 SVI 接口默认处于关闭状态，请读者手动开启接口
SW1(config-if)#int vlan 10
```

```
SW1(config-if)#no shutdown
SW1(config-if)#ip address 10.1.10.46 255.255.255.240
SW1(config)#no ip cef   //关闭思科模拟器交换设备设备上的"思科快速转发"功能，否则会造成
VLAN 间的数据不被转发，此处是多数思科模拟器的一个 bug
SW1#show ip int brief   //验证设备上配置 SVI 之后的 IP 地址状态
Interface              IP-Address      OK? Method Status        Protocol
Vlan8                  10.1.10.14      YES manual up            up
Vlan10                 10.1.10.46      YES manual up            up
//接口工作需要满足如下条件之一：① Trunk 允许对应 VLAN 流量通过；② 已经有接口被划入对
应 VLAN
SW1#show ip route
Codes: L - local, C - connected, S - static, R - RIP, M - mobile, B - BGP
       D - EIGRP, EX - EIGRP external, O - OSPF, IA - OSPF inter area
       N1 - OSPF NSSA external type 1, N2 - OSPF NSSA external type 2
       E1 - OSPF external type 1, E2 - OSPF external type 2
       i - IS-IS, su - IS-IS summary, L1 - IS-IS level-1, L2 - IS-IS level-2
       ia - IS-IS inter area, * - candidate default, U - per-user static route
       o - ODR, P - periodic downloaded static route, H - NHRP, l - LISP
       a - application route
       + - replicated route, % - next hop override

Gateway of last resort is not set

      10.0.0.0/8 is variably subnetted, 6 subnets, 3 masks
C        10.1.10.0/28 is directly connected, Vlan8    //SW1 上的 VLAN8 产生的直连路由
L        10.1.10.14/32 is directly connected, Vlan8
C        10.1.10.32/28 is directly connected, Vlan10  //SW1 上的 VLAN10 产生的直连路由
L        10.1.10.46/32 is directly connected, Vlan10
```

22.3.3 配置路由器模拟的终端并测试数据通信

关于该部分的详细介绍读者可以参考案例 20，请自行开启接口并配置对应的 IP 地址及网关。

```
PC1(config)#no ip routing
PC1(config)#ip default-gateway 10.1.10.14
S1(config)#no ip routing
S1(config)#ip default-gateway 10.1.10.46
!
PC1(config)#no ip routing
PC1(config)#ip default-gateway 10.1.10.14
PC1#ping 10.1.10.46   //PC1 可以和其他终端的网关通信，这意味着三层交换机工作正常
Type escape sequence to abort.
Sending 5, 100-byte ICMP Echos to 10.1.10.46, timeout is 2 seconds:
!!!!!
Success rate is 100 percent (5/5), round-trip min/avg/max = 1/202/1009 ms
```

PC1#ping 10.1.10.33   //PC1 和 S1 完成最终的数据通信
Type escape sequence to abort.
Sending 5, 100-byte ICMP Echos to 10.1.10.33, timeout is 2 seconds:
!!!!!

只有在完成本实验之后，才可以完成后续的路由技术、VRRP 技术等的学习，所以请熟练掌握本部分内容。读者需要自行完成后续案例中多个交换机的 SVI 配置。

本案例实施完毕。

案例 23　二层交换接口转换为三层路由接口的方案

交换机三层接口简书

设备的以太网接口工作在二层模式（很多初学者会在华为的以太网接口上配置三层的 IP 地址，这是不对的，以太网接口工作在二层，意味着其不能配置三层的 IP 地址，图 23-1 所示为初学者在二层接口上配置三层地址的常见错误），如果需要应用接口的三层功能，则需要使用特定命令将接口转换为三层模式。这种方式的扩展性较差，它会让交换机的一个物理接口独占一个广播域，所以一般不推荐使用（但最新的思科 SDN 企业网解决方案采用了该方案）。

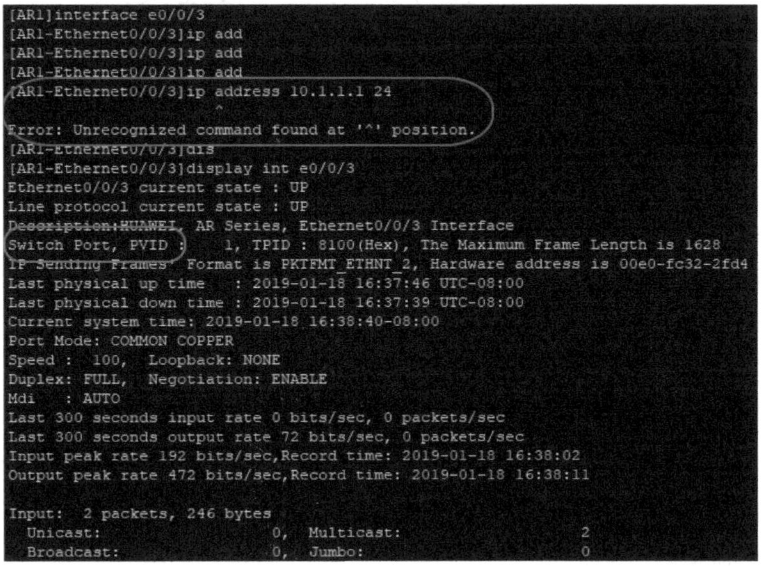

```
[AR1]interface e0/0/3
[AR1-Ethernet0/0/3]ip add
[AR1-Ethernet0/0/3]ip add
[AR1-Ethernet0/0/3]ip add
[AR1-Ethernet0/0/3]ip address 10.1.1.1 24
                        ^
Error: Unrecognized command found at '^' position.
[AR1-Ethernet0/0/3]dis
[AR1-Ethernet0/0/3]display int e0/0/3
Ethernet0/0/3 current state : UP
Line protocol current state : UP
Description:HUAWEI, AR Series, Ethernet0/0/3 Interface
Switch Port, PVID :    1, TPID : 8100(Hex), The Maximum Frame Length is 1628
IP Sending Frames' Format is PKTFMT_ETHNT_2, Hardware address is 00e0-fc32-2fd4
Last physical up time   : 2019-01-18 16:37:46 UTC-08:00
Last physical down time : 2019-01-18 16:37:39 UTC-08:00
Current system time: 2019-01-18 16:38:40-08:00
Port Mode: COMMON COPPER
Speed :  100,  Loopback: NONE
Duplex: FULL,  Negotiation: ENABLE
Mdi   : AUTO
Last 300 seconds input rate 0 bits/sec, 0 packets/sec
Last 300 seconds output rate 72 bits/sec, 0 packets/sec
Input peak rate 192 bits/sec,Record time: 2019-01-18 16:38:02
Output peak rate 472 bits/sec,Record time: 2019-01-18 16:38:11

Input:  2 packets, 246 bytes
  Unicast:                  0,  Multicast:                  2
  Broadcast:                0,  Jumbo:                      0
```

图 23-1　初学者在二层接口上配置三层地址的常见错误

23.1　交换机的三层路由接口实施案例

交换机的三层路由接口实施拓扑如图 23-2 所示。在图 23-2 中，把 SW2 的二层接口转换为一个具有路由功能的三层接口。通过把二层接口转换为三层接口实现 VLAN 间数据通信，注意，并不是所有系统或设备都支持该功能，尤其是在华为的 eNSP 模拟器上，命令可以支持，但功能却无法实现。

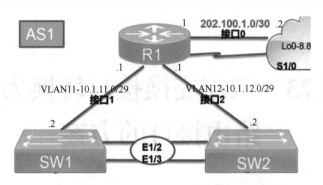

图 23-2 交换机的三层路由接口实施拓扑

23.2 交换机的三层路由接口实施拓扑

在图 23-2 中的 SW2 上，把接口 2 从默认的二层交换接口修改为三层路由接口，地址配置为 10.1.12.2/29，以便于和 R1 的接口 1（地址为 10.1.12.1/29）通信，这也是在多层交换机上实现三层直连网络通信的一种方式。

23.3 交换机三层接口配置要点

尽量避免把交换机上的二层接口改为三层接口，因为交换机的作用就是在"共享"的基础上实施管理，而不是让某个接口专门"独享"某个 IP 网络，通常情况下三层接口并不具备太好的扩展性。交换机三层接口配置命令如下：
① 思科设备的命令为 no switchport。
② 华为设备的命令为 undo portswitch。
③ 华三设备的命令为 port link-mode route。

23.4 交换机三层接口配置步骤

23.4.1 在思科设备上配置交换机三层接口

请在图 23-2 中 SW2 的 e0/2（图中接口 2）
上执行以下命令。

```
SW2(config)#int e0/2
SW2(config-if)#no switchport
SW2(config-if)#ip add
SW2(config-if)#ip address 10.1.12.2 255.255.255.248
```
请自行配置 R1 的接口地址，然后进行数据测试。
```
SW2#ping 10.1.12.1    //测试 SW2 连接的 R1 的地址，实现通信
Type escape sequence to abort.
Sending 5, 100-byte ICMP Echos to 10.1.12.1, timeout is 2 seconds:
.!!!!
Success rate is 80 percent (4/5), round-trip min/avg/max = 1/1/1 ms
```

测试完毕，结果显示实现了交换机和路由器的三层直连通信。

23.4.2 在华为设备上实施交换机三层接口

华为设备的 5700 系列和 6700 系列交换机中仅 S5720EI、S5720HI、S5730HI、S6720EI、S6720HI 和 S6720S-EI 支持二层模式与三层模式转换。

工作在三层模式的以太网接口支持配置 IP 地址，配置实例如下：

```
<HUAWEI> system-view
[HUAWEI] interface gigabitethernet 0/0/1
[HUAWEI-GigabitEthernet0/0/1] undo portswitch
[HUAWEI-GigabitEthernet0/0/1] ip address 10.1.12.2 255.255.255.248
```

请读者注意，eNSP 模拟器并不支持该配置（有些版本的模拟器可能可以支持该命令，但依旧无法实现相应功能）。

本案例实施完毕。

案例 24　在华为设备上配置 VRRP

VRRP 简书

主机将发送给外部网络的报文发送给网关，由网关传递给外部网络，从而实现主机与外部网络的通信。在正常的情况下，主机可以完全信赖网关的工作，但是当网关出现故障时，主机与外部的通信就会中断。VRRP（虚拟路由冗余协议）既不需要改变现有组网情况，也不需要在主机上做任何配置，只需要在相关路由器（交换机上）上配置 VRRP，就能实现下一跳网关的备份，并且不会给主机带来任何负担。思科设备除了 VRRP 还具备 HSRP 和 GLBP 这些类似的功能性三层协议，这些技术其实实现的是协议层次的网关冗余，而之前讲过的堆叠技术则是在硬件层次提高网络的可靠性。VRRP 示意图如图 24-1 所示，图中交换机 Switch A、Switch B、Switch C 配置了 VRRP，在用户侧是感知不到的，用户只知道网关 Switch X 而已，数据通过 VRRP 中的 Master 设备，即 Switch A 转发，当 Switch A 出现问题时，流量可以切换到其他备份设备。

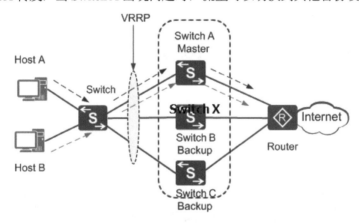

图 24-1　VRRP 示意图

表 24-1 为 VRRP 术语表，该表给出了关于 VRRP 的一些重要概念。

表 24-1　VRRP 术语表

术语	功能
虚拟路由器	由一个 Master 路由器和多个 Backup 路由器组成。主机将虚拟路由器当作默认网关
VRID/Group ID	虚拟路由器的标识。有相同 VRID 的一组路由器构成一台虚拟路由器
Master 路由器	虚拟路由器中承担报文转发任务的路由器
Backup 路由器	当 Master 路由器出现故障时，能够代替 Master 路由器工作的路由器
虚拟 IP 地址	虚拟路由器的 IP 地址。一台虚拟路由器可以拥有一个或多个 IP 地址

(续表)

术 语	功 能
IP 地址拥有者	接口 IP 地址与虚拟 IP 地址相同的路由器被称为 IP 地址拥有者，其优先级最大且不可修改
虚拟 MAC 地址	一台虚拟路由器拥有一个虚拟 MAC 地址。虚拟 MAC 地址的格式为 00-00-5E-00-01-{VRID}。通常情况下，虚拟路由器回应 ARP 请求使用的是虚拟 MAC 地址，只有对虚拟路由器进行了特殊配置，才回应接口的真实 MAC 地址
优先级	VRRP 根据优先级来确定虚拟路由器中每台路由器的地位，数值越大优先级越高
抢占方式	如果 Backup 路由器工作在抢占方式下，当它收到 VRRP 报文后，会将自己的优先级与通告报文中的优先级进行比较。如果自己的优先级比当前的 Master 路由器的优先级高，就会主动抢占成为 Master 路由器；否则，将保持 Backup 状态

（1）VRRP 的工作过程

① 虚拟路由器中的路由器根据优先级选举出 Master 路由器。Master 路由器通过发送免费 ARP 报文，将自己的虚拟 MAC 地址告知与它连接的设备或者主机，从而承担报文转发任务。

② Master 路由器周期性地发送 VRRP 报文，以公布其配置信息（优先级等）和工作状况。

③ 如果 Master 路由器出现故障，虚拟路由器中的 Backup 路由器将根据优先级重新选举新的 Master。

④ 当虚拟路由器切换状态时，Master 路由器由一台设备切换为另外一台设备，新的 Master 路由器只是简单地发送一个携带虚拟路由器的 MAC 地址和虚拟 IP 地址信息的免费 ARP 报文，这样就可以更新与它连接的主机或设备中的 ARP 相关信息。而网络中的主机感知不到 Master 路由器已经切换为另外一台设备。

⑤ 当 Backup 路由器的优先级高于 Master 路由器时，由 Backup 路由器的工作方式（抢占方式和非抢占方式）决定是否重新选举 Master 路由器。

由此可见，为了保证 Master 路由器和 Backup 路由器能够协调工作，VRRP 需要实现以下功能：

- Master 路由器的选举；
- Master 路由器状态的通告；
- 为了提高安全性，VRRP 还需提供认证功能。

（2）Master 路由器的选举

VRRP 根据优先级来确定虚拟路由器中每台路由器的角色（Master 路由器或 Backup 路由器）。优先级越高，则越有可能成为 Master 路由器。

初始时路由器工作在 Backup 状态，通过 VRRP 报文的交互获知虚拟路由器中其他成员的优先级：

- 如果 VRRP 报文中 Master 路由器的优先级高于自己的优先级，则路由器保持 Backup 状态；
- 如果 VRRP 报文中 Master 路由器的优先级低于自己的优先级，采用抢占工作方式的路由器将抢占成为 Master 路由器，周期性地发送 VRRP 报文，采用非抢占工作方式的路由器仍保持 Backup 状态；
- 如果在一定时间内没有收到 VRRP 报文，则路由器切换为 Master 状态。

VRRP 优先级的取值范围为 0～255（数值越大表明优先级越高），可配置的范围是 1～254，优先级 0 是系统保留给路由器当其放弃 Master 位置时使用的，255 则是系统保留给 IP 地址拥有者使用的。当路由器为 IP 地址拥有者时，其优先级始终为 255。因此，当虚拟路由器中存在 IP 地址拥有者时，只要其工作正常，则为 Master 路由器。

（3）监视上行链路

VRRP 网络传输功能有时需要靠额外的技术来完善其工作。例如，当 Master 路由器到达某网络的链路突然断掉时，主机无法通过此 Master 路由器远程访问该网络。此时，可以通过监视指定接口上行链路功能来解决这个问题。当 Master 路由器发现上行链路出现故障后，主动降低自己的优先级（使 Master 路由器的优先级低于 Backup 路由器），并立即发送 VRRP 报文。Backup 路由器接收到优先级比自己低的 VRRP 报文后，等待 Skew Time 切换为新的 Master 路由器。从而使得能够到达此网络的 Backup 路由器充当 VRRP 新的 Master 路由器，协助主机完成网络通信。

VRRP 可以直接监视连接上行链路的接口状态。当连接上行链路的接口关闭时，将 Master 路由器降低指定的优先级。VRRP 的优先级最低可以降低到 1。

VRRP 可以利用 NQA/IP SLA 技术监视上行链路连接的远端主机或网络状况。例如，在 Master 设备上启动 NQA/IP SLA 的 ICMP Echo 探测功能，探测远端主机的可达性。当 ICMP Echo 探测失败时，它可以通知本设备探测结果，达到降低 VRRP 优先级的目的。

VRRP 也可以利用 BFD 技术监视上行链路连接的远端主机或网络状况。由于 BFD 的精度可以到达 10 ms，通过 BFD 能够快速检测到链路状态的变化，从而达到快速抢占的目的。例如，可以在 Master 路由器上使用 BFD 技术监视上行设备的物理状态，在上行设备发生故障后，快速检测到该变化，并降低 Master 路由器的优先级，致使 Backup 路由器等待 Skew Time 后，抢占成为新的 Master 路由器。

24.1　VRRP 实施案例

在汇聚交换机 SW1 和 SW2 中实施 VRRP 以保证终端网关设备的高可用性（当某个网关设备失效后，其他网关设备依旧可以实现业务数据的转发）。我们将在图 24-2 中展示该功能，对应三个业务 VLAN 的网关地址（即虚拟 IP 地址）为 10.1.10.14、10.1.10.30 和 10.1.10.46。

24.2　VRRP 案例拓扑

在汇聚层设备实施 VRRP 如图 24-2 所示，图中的 SW1 和 SW2 上配置了 VRRP。涉及的业务 VLAN 为 VLAN8、VLAN9 和 VLAN10。请在核心交换机 SW1 和 SW2 上配置物理地址后再配置虚拟网关地址。

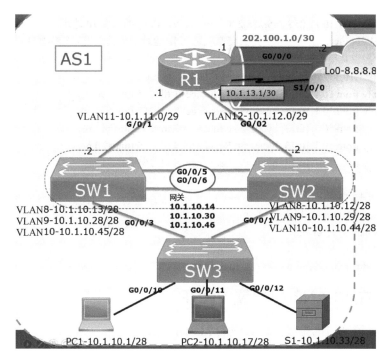

图 24-2 在汇聚层设备实施 VRRP

24.3 VRRP 配置要点

① 正确完成 VLAN 的基础配置（接入和 Trunk 模式）。
② 正确完成 VLANIF 的配置（同一网络）。
③ 配置一致的 VRRP 的虚拟 IP 地址（同一网络）。
④ 配置优先级管理 Master 设备。

24.4 配置 VRRP 步骤详解

配置 VRRP 的前置条件：请确保交换机之间的 Trunk 配置及连接终端接口的接入配置正确，这样可以确保 VLANIF 正常工作。

24.4.1 在交换机上配置直连的 VLANIF 地址

在交换机上配置 3 个业务 VLAN 的真实 IP 地址，此处的"真实 IP 地址"对应后续 VRRP 的虚拟 IP 地址，此处以 VLAN8 为例，读者可以参照拓扑自行配置另外 2 个 VLAN。

配置 SW1：
```
interface Vlan8
 ip address 10.1.10.13 255.255.255.240   //配置 SW1 设备 VLANIF 的真实 IP 地址，该地址偶尔被
工程师称为物理地址
```
配置 SW2：
```
interface Vlan8
```

ip address 10.1.10.12 255.255.255.240   //配置 VLANIF 的真实 IP 地址
<HW-SW2>ping 10.1.10.13   //测试交换机之间 IP 地址的通信情况，VRRP 并不是二层协议，而是 IP 协议，所以这一步也是需要的过程
  ping 10.1.10.13: 56   data bytes, press CTRL_C to break
  Reply from 10.1.10.13: bytes=56 Sequence=1 ttl=255 time=30 ms
  Reply from 10.1.10.13: bytes=56 Sequence=2 ttl=255 time=30 ms
  Reply from 10.1.10.13: bytes=56 Sequence=3 ttl=255 time=1 ms
  Reply from 10.1.10.13: bytes=56 Sequence=4 ttl=255 time=20 ms
  Reply from 10.1.10.13: bytes=56 Sequence=5 ttl=255 time=1 ms

  --- 10.1.10.13 ping statistics ---
  5 packet(s) transmitted
  5 packet(s) received
  0.00% packet loss
  round-trip min/avg/max = 1/16/30 ms

24.4.2　在 VLAN8 中配置 VRRP

为 SW1 配置较高的优先级，使其成为 Master 设备。

SW1：
interface Vlanif8
 ip address 10.1.10.13 255.255.255.240
 vrrp vrid 8 virtual-ip 10.1.10.14   //配置 VRRP 组 8，并配置虚拟 IP 地址，注意，该 IP 地址才是该 VLAN 中终端设备的网关地址
 vrrp vrid 8 priority 110   //配置 SW1 的优先级，默认为 100，此处修改为 110
SW2：
interface Vlanif8
 ip address 10.1.10.12 255.255.255.240
 vrrp vrid 8 virtual-ip 10.1.10.14   //配置 VRRP 组 8，并配置虚拟 IP 地址，该虚拟 IP 地址和 SW1 上的配置相同

验证配置后的结果：
<HW-SW2>display vrrp 8
  Vlanif8 | Virtual Router 8
    State : Slave   //SW2 为 Slave 状态
    Virtual IP : 10.1.10.14   //配置虚拟 IP 地址
    Master IP : 10.1.10.13   //主设备（即 SW1 设备）的 IP 地址是 10.1.10.13
    PriorityRun : 100
    PriorityConfig : 100   //配置的优先级为默认的 100
    MasterPriority : 110   //主设备的优先级
    Preempt : YES   Delay Time : 0 s
    TimerRun : 1 s
    TimerConfig : 1 s   //通告间隔，默认为 1 s
    Auth type : NONE
    Virtual MAC : 0000-5e00-0108   //通过组号得到的虚拟 MAC 地址，主设备会使用该 MAC 地址响应终端的 ARP 请求
    Check TTL : YES
    Config type : normal-vrrp

```
            Create time : 2018-12-28 15:19:54 UTC-08:00
            Last change time : 2018-12-28 15:20:10 UTC-08:00
```
测试终端和网关的通信情况：
```
    PC>ping 10.1.10.14

    ping 10.1.10.14: 32 data bytes, Press Ctrl_C to break
    From 10.1.10.14: bytes=32 seq=1 ttl=255 time=78 ms
    From 10.1.10.14: bytes=32 seq=2 ttl=255 time=32 ms
    From 10.1.10.14: bytes=32 seq=3 ttl=255 time=31 ms
    From 10.1.10.14: bytes=32 seq=4 ttl=255 time=31 ms
    From 10.1.10.14: bytes=32 seq=5 ttl=255 time=31 ms

    --- 10.1.10.14 ping statistics ---
      5 packet(s) transmitted
      5 packet(s) received
      0.00% packet loss
      round-trip min/avg/max = 31/40/78 ms

    PC>
```
查看 PC 设备上的 arp 表项，可以看到网关 IP 地址对应一个虚拟 MAC 地址。
```
    PC>arp -a

    Internet Address     Physical Address       Type
    10.1.10.13           4C-1F-CC-B7-3B-11      dynamic
    10.1.10.14           00-00-5E-00-01-08      dynamic    //该 MAC 地址正是 VRRP 的 Master 设备做
出响应的虚拟 MAC 地址
```

24.4.3 在其他 VLAN 中配置 VRRP

在本节中将配置 SW1 为 VLAN10 的 Master 设备，SW2 为 VLAN10 的 Slave 设备；为了保证一定程度的流量负载，SW2 会成为 VLAN9 的 Master 设备，SW1 会成为 VLAN9 的 Slave 设备。读者请注意，STP 根设备应该和 VRRP 的 Master 设备是同一台交换机，否则会出现二层次优转发情况。

配置 SW1：
```
    interface Vlanif9
     ip address 10.1.10.28 255.255.255.240
     vrrp vrid 9 virtual-ip 10.1.10.30
    interface Vlanif10
     ip address 10.1.10.45 255.255.255.240
     vrrp vrid 10 virtual-ip 10.1.10.46
     vrrp vrid 10 priority 110
```
配置 SW2：
```
    interface Vlanif9
     ip address 10.1.10.29 255.255.255.240
     vrrp vrid 9 virtual-ip 10.1.10.30
     vrrp vrid 9 priority 110
```

```
interface Vlanif10
 ip address 10.1.10.44 255.255.255.240
 vrrp vrid 10 virtual-ip 10.1.10.46
```

验证交换机上 VRRP 的状态：

```
<HW-SW1>dis vrrp brief
VRID    State       Interface       Type        Virtual IP
----------------------------------------------------------------
8       Master      Vlanif8         Normal      10.1.10.14
9       Backup      Vlanif9         Normal      10.1.10.30
10      Master      Vlanif10        Normal      10.1.10.46
----------------------------------------------------------------
Total:3    Master:2    Backup:1    Non-active:0

[HW-SW2]dis vrrp brief
VRID    State       Interface       Type        Virtual IP
----------------------------------------------------------------
8       Backup      Vlanif8         Normal      10.1.10.14
9       Master      Vlanif9         Normal      10.1.10.30
10      Backup      Vlanif10        Normal      10.1.10.46
----------------------------------------------------------------
Total:3    Master:1    Backup:2    Non-active:0
```

设备终端通信验证和虚拟 MAC 地址如图 24-3 所示。

图 24-3　设备终端通信验证和虚拟 MAC 地址

华三设备与华为设备配置 VRRP 的命令基本相同，如下所示。

```
vrrp vrid X virtual-ip virtual-address    //配置一个虚拟 IP 地址
vrrp vrid X priority 110    //配置备份组的优先级为 110
```

本案例实施完毕。

案例 25　在思科设备上配置 VRRP

25.1　VRRP 实施案例及拓扑

在思科设备上配置 VRRP，如图 25-1 所示。在汇聚交换机（SW1 和 SW2）上配置 VRRP 以保证终端网关设备的高可用性，各个业务 VLAN 的 SVI 地址如图 25-1 所示，这些地址即真实地址。VLAN8、VLAN9 和 VLAN10 的网关虚拟 IP 地址分别为 10.1.10.14、10.1.10.30 和 10.1.10.46。

请注意，要使对应 VLAN 的生成树根设备和 VRRP 的转发设备位于同一网络，从而避免二层数据帧转发的次优路径。

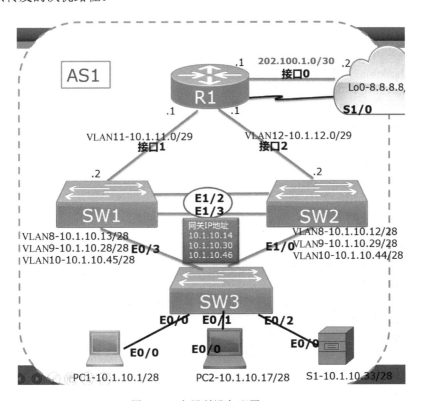

图 25-1　在思科设备配置 VRRP

25.2　VRRP 配置要点

① 正确配置 VLAN。
② 正确配置 SVI。

③ 配置 VRRP 的虚拟 IP 地址（同一网络）。
④ 配置优先级管理 Master 设备。

25.3 VRRP 配置步骤

配置 VRRP 的前置条件：请确保交换机之间的 Trunk 正确，连接终端接口接入配置正确，否则如果 SVI 不工作，则 VRRP 无法实施。

配置 SW1：
```
interface Vlan8
 no shutdown
 ip address 10.1.10.13 255.255.255.240
 vrrp 8 ip 10.1.10.14    //配置VRRP 组8，使用组8 是为了方便标识针对 VLAN8 的配置，虚拟IP
地址 10.1.10.14 是终端的网关地址
 vrrp 8 priority 110    //配置较大的优先级使 SW1 成为 VLAN8 的主设备
interface Vlan9
 no shutdown
 ip address 10.1.10.28 255.255.255.240
 vrrp 9 ip 10.1.10.30
interface Vlan10
 no shutdown
 ip address 10.1.10.45 255.255.255.240
 vrrp 10 ip 10.1.10.46
 vrrp 10 priority 110
```

配置 SW2：
```
interface Vlan8
 no shutdown
 ip address 10.1.10.12 255.255.255.240
 vrrp 8 ip 10.1.10.14
interface Vlan9
 no shutdown
 ip address 10.1.10.29 255.255.255.240
 vrrp 9 ip 10.1.10.30
 vrrp 9 priority 110    //为 SW2 配置较高的优先级使其成为 VLAN9 的主设备
interface Vlan10
 no shutdown
 ip address 10.1.10.44 255.255.255.240
 vrrp 10 ip 10.1.10.46
```

验证 VRRP 配置，针对不同 VLAN 的不同角色，可以看到 SW1 是 VLAN8 和 VLAN10 的主设备，而 SW2 是 VLAN9 的主设备。

```
C-SW1#show vrrp brief
Interface  Grp  Pri  Time  Own  Pre State  Master addr  Group addr
Vl8        8    110  3570  Y    Master     10.1.10.13   10.1.10.14
Vl9        9    100  3609  Y    Backup     10.1.10.29   10.1.10.30
Vl10       10   110  3570  Y    Master     10.1.10.45   10.1.10.46
C-SW2#show vrrp brief
```

Interface	Grp	Pri	Time	Own	Pre State	Master addr	Group addr
Vl8	8	100	3609	Y	Backup	10.1.10.13	10.1.10.14
Vl9	9	110	3570	Y	Master	10.1.10.29	10.1.10.30
Vl10	10	100	3609	Y	Backup	10.1.10.45	10.1.10.46

测试终端与网关的数据通信情况：

```
C-PC1#ping 10.1.10.14    //终端设备完成与网关设备的通信
Type escape sequence to abort.
Sending 5, 100-byte ICMP Echos to 10.1.10.14, timeout is 2 seconds:
!!!!!
Success rate is 100 percent (5/5), round-trip min/avg/max = 1/202/1010 ms
C-PC1#show arp
Protocol   Address       Age (min)  Hardware Addr    Type   Interface
Internet   10.1.10.1     -          aabb.cc00.8000   ARPA   Ethernet0/0
Internet   10.1.10.14    0          0000.5e00.0108   ARPA   Ethernet0/0   //PC 得到 ARP 表项,
```
可以看到其中的 MAC 地址是由 VRRP 组（08）映射得到的

关于 VRRP 的其他一些配置还包括认证、跟踪条件以完成状态切换等，这部分内容请关注笔者的其他作品。

本案例实施完毕。

案例 26　在华为设备上实施端口安全技术

交换机端口安全技术简书

端口安全是一种交换机安全技术，其在一定程度上可以保证企业交换网络的安全，它的本质是通过限制 MAC 地址表实现对网络的管控。一般情况下可用于限制接入端口下的授权用户，非授权用户或者过多的用户接入会引发安全处理行为（比如限制用户私自接入家用级别路由器）。如果应用在接入层设备中，通过配置端口安全可以防止仿冒用户从其他端口攻击。如果应用在汇聚层设备中，通过配置端口安全可以控制接入用户的数量。

在开启端口安全的默认情况下，每个端口仅可以学习一个安全 MAC 地址，用户可以配置端口学习安全 MAC 地址的最大数量限制。用户也可以配置端口的安全保护动作，当端口学习到的安全 MAC 地址数量达到限制时，可以选择采取下列某一种动作。

- protect：当学习到的 MAC 地址数达到端口限制数时，端口丢弃源地址在 MAC 表以外的报文。
- restrict：当学习到的 MAC 地址数超过端口限制数时，端口丢弃源地址在 MAC 表以外的报文，同时发出告警。
- shutdown：当学习到的 MAC 地址数超过端口限制数时，端口执行 shutdown 操作，同时发出告警。

默认情况下，华为设备的端口安全保护动作为 restrict，思科设备的端口安全保护动作为 shutdown。

26.1　实施端口安全技术的拓扑

在接入层交换机实施端口安全技术可以确保非法用户无法接入到企业内部网络，同时还可以控制终端用户的接入数量。

图 26-1 为在 SW3 上实施端口安全技术的拓扑，实施该技术涉及的设备较少，实验较易完成。在进行本步骤之前，请实现基本的 VLAN 间数据通信，这样更便于后续数据测试。

26.2　端口安全的配置要点

① 完成交换机 VLAN 的端口接入配置。
② 在连接用户接口处开启端口安全。
③ 验证端口安全的违规行为，掌握如何处理违规端口的方法。

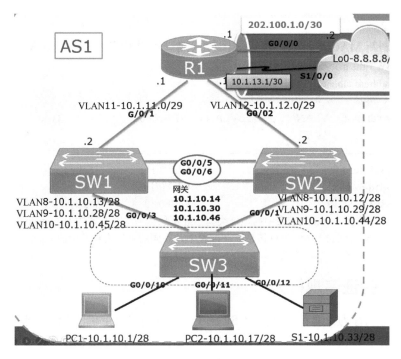

图 26-1 在 SW3 上实施端口安全技术的拓扑

26.3 配置端口安全步骤详解

26.3.1 在交换机连接终端的端口上实施端口安全技术

此处进行了两方面的修改：一是开启端口安全命令；二是将端口安全开启后该端口学习 MAC 地址的方式变为"sticky"方式。

```
SW3:
Interface GigabitEthernet0/0/10
  port link-type access
  port default vlan 8
  port-security enable        //启动端口安全技术
  port-security mac-address sticky   //命令配置端口安全功能后，学习到的 MAC 地址变为安全动态 MAC
地址。而使能 sticky MAC 功能后端口会将学习到的安全动态 MAC 地址转化为 sticky MAC 地址，即
  interface GigabitEthernet0/0/11
  port link-type access
  port default vlan 9
  port-security enable
  port-security mac-address sticky
  interface GigabitEthernet0/0/12
  port link-type access
  port default vlan 10
  port-security enable
  port-security mac-address sticky
```

对于连续端口，读者也可以采用如下命令：port-group group-member Gigabit Ethernet

0/0/10 to g0/0/12。

为了尽快刷新 MAC 地址表，可以关闭接入端口后再开启，然后验证 MAC 地址表。

```
[HW-SW3]dis mac-address sticky
MAC address table of slot 0:
---------------------------------------------------------------
MAC Address      VLAN/        PEVLAN CEVLAN Port       Type       LSP/LSR-ID
                 VSI/SI                                           MAC-Tunnel

5489-9841-1c54   9            -      -      GE0/0/11   sticky     -
5489-985d-62d4   8            -      -      GE0/0/10   sticky     -
5489-9883-7156   10           -      -      GE0/0/12   sticky     -
//读者可以看到 MAC 地址表中的地址类型已经从默认的动态方式变为了 sticky 方式，在大部分网
络产品中，sticky 指学习到的 MAC 地址被粘贴到配置中，不同的是华为设备中的 sticky 方式并不体现在配
置中
```

26.3.2 验证违规行为

为了模拟违规行为，读者可以在 eNSP 的 PC 端手动修改主机的 MAC 地址，这样交换机会从运行端口安全功能的端口收到两个 MAC 地址，而默认情况下运行端口安全功能的端口学习到的 MAC 地址数最多为 1，从而产生违规行为，修改终端的 MAC 地址如图 26-2 所示。

图 26-2　修改终端的 MAC 地址

```
          Nov 30 2018 16:49:03-08:00 HW-SW3 L2IFPPI/4/PORTSEC_ACTION_ALARM:OID 1.3.6.1.4.1.
2011.5.25.42.2.1.7.6 The number of MAC address on interface (16/16) GigabitEthernet0/0/10 reaches the limit, and
the port status is : 1. (1:restrict;2:protect;3:shutdown)    //该日志表明，G0/0/10 口由于学习到的 MAC 地址到达上
限，所以端口状态被置为 restrict 状态，即丢弃该流量，发送日志，端口 up，统计违规行为数
```

请读者在 PC1 上修改回正确的 MAC 地址。如果要使因违规而关闭的端口重新开启，可以先手动关闭端口再开启该端口；或者使用自动恢复方式，命令为"error-down auto-recovery portsec-reachedlimit"，但 eNSP 并不支持该命令。另外，需要注意，有时在 eNSP 上修改终端的 MAC 地址验证效果并不完美。

26.3.3 华三设备实施端口安全参考命令

华三设备实施端口安全参考命令如下，其他命令读者可以通过键入"？"寻求帮助。

```
[Device]port-security enable     //开启端口安全
[Device] interface ten-gigabitethernet 1/0/1
```

[Device-Ten-GigabitEthernet1/0/1] port-security max-mac-count 2　　//设置端口安全允许的最大安全MAC 地址数为 2
　　[Device-Ten-GigabitEthernet1/0/1] port-security port-mode autolearn　　//设置端口安全模式为 autolearn
　　[Device-Ten-GigabitEthernet1/0/1] port-security intrusion-mode disableport-temporarily　　//设置触发入侵检测特性后的保护动作为暂时关闭端口
　　[Device-Ten-GigabitEthernet1/0/1] quit
　　[Device] port-security timer disableport 30　　//关闭时间为 30 秒

本案例实施完毕。

案例 27　在思科设备上实施端口安全技术

27.1　实施端口安全技术的拓扑

在接入层交换机上实施端口安全技术可以确保非法用户无法接入企业内部网络。

图 27-1 所示为在 SW3 上实施端口安全技术的拓扑。在完成本步骤之前，请实现基本的 VLAN 间数据通信，这样可以方便后续测试。

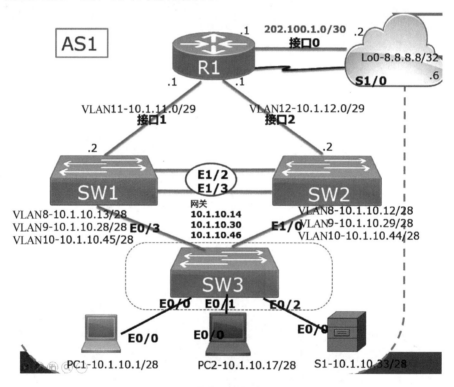

图 27-1　在 SW3 上实施端口安全技术的拓扑

27.2　端口安全的配置要点

① 完成交换机 VLAN 的端口接入配置。
② 在连接用户的接口上开启端口安全功能。
③ 验证端口安全的违规行为，掌握如何处理违规端口的方法。

27.3 配置端口安全步骤详解

27.3.1 在交换机上实施端口安全技术

在实施端口安全技术之前，请确保端口已经成功划分到了正确的 VLAN 中。

```
SW3(config-if-range)#int range e0/0 - 2    //批量进入 e0/0 - 2 口，这些端口是连接终端的端口
SW3(config-if-range)#switchport port-security    //接口开启端口安全，这些接口默认只能学习到 1 个 MAC 地址
SW3(config-if-range)#switchport port-security mac-address sticky    //开启端口安全的"粘贴"MAC 地址功能，学习到的 MAC 地址将会被"粘贴"到配置中
```

注释：验证开启端口安全后的 MAC 地址表，请关注"Type"一列内容显示为"STATIC"，这点与华为不同（华为显示为 Sticky）。

验证设备学习到的 MAC 地址表（MAC 地址表是非常重要的参考标准）：

```
SW3#show mac address-table  static
          Mac Address Table
-------------------------------------------

Vlan    Mac Address        Type        Ports
----    -----------        --------    -----
 8      aabb.cc00.8000     STATIC      Et0/0
 9      aabb.cc00.9000     STATIC      Et0/1
10      aabb.cc00.b000     STATIC      Et0/2
```

SW3#show port-security   //验证端口安全的实施情况，可以看到配置了端口安全的端口默认情况下允许学习到的 MAC 地址数（默认数目为1），当前学习到的 MAC 地址数，违规行为统计（当然是 0），当发生违规行为时思科设备会默认关闭端口

Secure Port	MaxSecureAddr (Count)	CurrentAddr (Count)	SecurityViolation (Count)	Security Action
Et0/0	1	1	0	Shutdown
Et0/1	1	1	0	Shutdown
Et0/2	1	1	0	Shutdown

27.3.2 验证端口安全的违规行为和自动恢复功能

可以通过在终端上修改设备的 MAC 地址来增加交换机端口下学习到的 MAC 地址，当学习到的 MAC 地址数大于 1 时，则会违规。

```
C-PC1(config-if)#mac-address 1.1.1    //修改 PC1 物理端口的 MAC 地址
*Nov 30 13:07:26.825: %PM-4-ERR_DISABLE: psecure-violation error detected on Et0/0, putting Et0/0 in err-disable state    //在 Et0/0 端口上检测到违规行为，思科设备会默认关闭该端口
SW3#*Nov 30 13:07:26.825: %PORT_SECURITY-2-PSECURE_VIOLATION: Security violation occurred, caused by MAC address 0001.0001.0001 on port Ethernet0/0.
*Nov 30 13:07:27.832: %LINEPROTO-5-UPDOWN: Line protocol on Interface Ethernet0/0, changed state to down
```

```
SW3#
*Nov 30 13:07:28.827: %LINK-3-UPDOWN: Interface Ethernet0/0, changed state to down
SW3# show int e0/0，Ethernet0/0 is down, line protocol is down (err-disabled)    //查验交换机的日志，可以看到端口由于违规而被关闭
```

思科设备提供了自动和手动恢复接口的功能，自动恢复端口的功能使得管理员不必每次都手动关闭，然后再开启已经不再违规的端口。自动恢复端口的功能配置如下：

```
SW3(config)#errdisable recovery cause all    //开启差错自动恢复功能，此处选择了所有由于差错而关闭端口的原因
SW3(config)#errdisable recovery interval 30   //如果没有后续的违规行为将每隔30 s尝试一次，使端口恢复正常
C-PC1(config)#int e0/0
C-PC1(config-if)#no mac-address    //在终端上去掉手动配置的MAC地址，这样就不再有违规设备非法接入网络
*Nov 30  09:10:33.413: %PM-4-ERR_RECOVER: Attempting to recover from psecure-violation err-disable state on Et0/0   changed state to up    //可以看到自动恢复功能已经使之前违规的Et0/0端口恢复正常
```

本案例实施完毕。

第三篇 路由网络

本篇包含了重要的路由知识，我们把网络的边界做了进一步的扩展，即如何把数据从本地网络传输到远端网络。路由协议负责构建路由表，路由器（或者多层交换机）根据路由表，通过转发引擎把数据发送到下一跳设备。本篇为读者展示了基础路由知识，这些内容包括：静态路由配置、OSPF 多区域基本配置、GRE 隧道配置、基本 EBGP 邻居和产生路由配置。路由知识错综复杂，关于更详细的内容，请关注笔者的直播课程或者其他书籍。

本篇内容包含案例 28~33。

案例 28　在华为设备上配置静态路由

路由是数据通信网络中最基本的要素。路由信息就是指导报文发送的路径信息，路由的过程就是报文转发的过程。路由器最基本的两个功能就是路由决策（即通过路由表决策数据以何种方式转发）和数据转发（即通过路由表转发数据）。

获取路由的方式分为静态方式和动态方式两种。采用动态方式获取路由的协议被称为动态路由协议，它又可分为距离矢量协议（包含 RIP、EIGRP 和 BGP）和链路状态协议（包含 OSPF 和 IS-IS）。

路由器优选转发信息库（FIB）的标准：

路由器在优选路由后，会将路由表中被激活的路由下发到 FIB 表中，当报文到达路由器时，会通过查找 FIB 表进行转发。

FIB 表中每条转发项都指明到达某网段或某主机的报文应通过路由器的哪个物理接口或逻辑接口发送，然后就可到达该路径的下一个路由器，或者不再经过别的路由器而传送到与该设备直接相连的网络中的目的主机。

FIB 表的匹配遵循最长匹配原则。在查找 FIB 表时，报文的目的地址和 FIB 中各表项的掩码进行按位"逻辑与"，若得到的地址符合 FIB 表项中的网络地址匹配原则，则最终选择一个最长匹配的 FIB 表项转发报文。主机路由是最精准的路由，而默认路由是最不精准的路由。请注意路由表负载均衡，即当到达一个目标网络有多个下一跳时，数据报文并不是按照 1∶1 的方式转发的，而是会根据数据流的散列结果将流量转发到某一特定链路，当散列值不同时会转发到不同的链路。

路由必须有直连的下一跳才能够转发报文，但在路由生成时下一跳可能不是直连的，因此需要计算出一个直连的下一跳和对应的出接口，这个过程就叫作路由迭代。BGP 路由、静态路由的下一跳都有可能不是直连的，需要进行路由迭代，其他 IGP（内部网关协议）的下一跳是直连网络。

路由原理和静态路由简书

路由表是数据转发的决策表，决定放入路由表中的属性有路由优先级和度量值。

如果到相同目的地址有多个路由来源（即多种路由协议竞争路由表），则以 Preference［优先级，是华为、华三的说法；Admin Distance（管理距离）是思科的说法］确定不同类型协议优先级，Preference 值越小，优先级越高，优先级最高的路由被添加进路由表。路由优先级/管理距离工作示意图如图 28-1 所示。各厂商协议的优先级不尽相同，协议在不同厂商设备上应用时的优先级如图 28-2 所示。

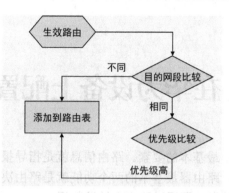

图 28-1　路由优先级／管理距离工作示意图

	思科	华为	Juniper
直连	0	0	0
静态	1	60	5
RIP	120	100	100
OSPF内部	110	10	10
OSPF外部	110	150	150
IS-IS	115	15	15（可以更细化）
iBGP	200	255	170
EBGP	20	255	150
EIGRP	内部90，外部170		

图 28-2　协议在不同厂商设备上应用时的优先级

度量值（Metric）是某一协议内部衡量到达某个目的地多跳路径的标准，度量值越小越优先。通常影响路由度量值的因素有：线路延迟、带宽、线路使用率、线路可信度、跳数、最大传输单元等。但其实除了 EIGRP 和 BGP，其他协议都采用单一度量方式，比如 RIP 采用跳数、OSPF 采用接口带宽、IS-IS 采用每个接口的开销数默认为 10 的方式。

IP 报文转发是一种"逐跳转发"的行为，即在沿途所有设备上都要进行 IP 路由方式处理，如果没有匹配路由，那么设备将丢弃 IP 报文。单个设备查找路由表（RIB）的过程如图 28-3 所示。

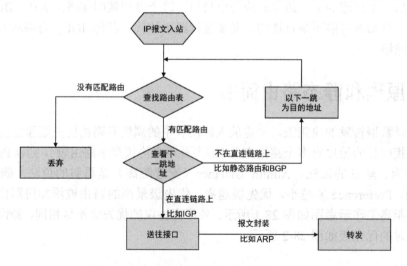

图 28-3　单个设备查找路由表（RIB）的过程

静态路由是一种需要管理员手动配置的特殊路由，现网中静态路由应用场景很多，因为它可以精准地控制数据转发，所以当网络结构比较简单时，只需要配置静态路由就可以使网络正常工作。在复杂网络环境中，配置静态路由可以改进网络的性能，并可为重要的应用保证带宽，但不能大面积使用。

静态路由配置简单、可控性高、使用带宽少，并且不占用 CPU 资源来计算和分析路由更新信息。但是当网络发生故障时或拓扑发生变化后，静态路由不会自动更新，必须重新手动配置。因此，静态路由不适用于大型和复杂的网络环境。一方面，网络管理员难以全面了解整个网络的拓扑结构；另一方面，当网络的拓扑结构或链路状态发生变化时，需要对路由器中的静态路由信息进行大范围地调整，此工作的难度大，复杂程度非常高。

在配置静态路由时，根据不同的出接口类型，指定出接口和下一跳地址。对于点到点类型的接口，只需指定出接口即可。因为指定发送接口即隐含指定了下一跳地址，这时认为与该接口相连的对端接口地址就是路由的下一跳地址。

对于广播类型的接口（如以太网接口），必须指定通过该接口发送数据时对应的下一跳地址。因为以太网接口是广播类型的接口，会导致出现多个下一跳地址，无法唯一确定下一跳地址，华为设备由于没有开启代理 ARP 会导致报文被丢弃，而在思科设备中可采用技术转发流量，但严重降低设备性能，在本书后续案例中会演示这种场景。

28.1 静态路由应用案例及拓扑

在企业网关设备上配置静态路由拓扑，如图 28-4 所示，图中展示了在企业网络的网关设备（R1）上配置静态路由、浮动静态路由和默认路由，从而保证数据可以转发到互联网。如果缺少路由或者存在错误的路由，设备会丢弃数据，无法完成数据的转发。

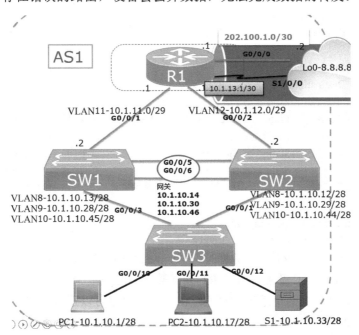

图 28-4　在企业网关设备上配置静态路由拓扑

28.2 配置静态路由要点

配置静态路由的要点如下：
① 在保证接口工作的前提下配置静态明细路由。
② 配置静态浮动路由，理解优先级的作用。
③ 配置默认路由以访问互联网，理解下一跳和出接口的异同。

28.3 配置静态路由步骤详解

28.3.1 在网关设备上配置静态明细路由

确保接口已经配置了正确的 IP 地址且可以正常工作，否则，如果接口不工作，那么静态路由将不会出现在路由表中。

[HW-R1]ip route-static 8.8.8.8 32 s1/0/0   //配置到达地址为 8.8.8.8/32 的网络的出接口为 s1/0/0，在点到点类型的网络上（包含 PPP 链路、HDLC 链路或者帧中继的点到点子接口），由于数据发送出去只能到达对端设备，所以不需要配置下一跳地址

[HW-R1]ip route-static 8.8.8.8 32 GigabitEthernet 0/0/0 202.100.1.2   //配置到达地址为 8.8.8.8/32 的网络的出接口为 G0/0/0，下一跳地址为 202.100.1.2，在以太网链路上，这种接口是一种多点接入网络，强烈推荐使用出接口和下一跳（202.100.1.2）的实施方案

验证配置静态路由之后，得到的静态路由表：

```
<HW-R1>display ip routing-table protocol static
Route Flags: R - relay, D - download to fib
------------------------------------------------------------
Public routing table : Static
        Destinations : 1      Routes : 2      Configured Routes : 2

Static routing table status : <Active>
        Destinations : 1      Routes : 2

Destination/Mask    Proto    Pre   Cost      Flags    NextHop          Interface

       8.8.8.8/32   Static   60    0         D        202.100.1.2      GigabitEthernet0/0/0
                    Static   60    0         D        12.1.1.1         Serial1/0/0  //到达目标
```
地址 8.8.8.8/32 的静态路由，出现负载均衡的情况，在华为设备中这些静态路由的优先级为 60

```
Static routing table status : <Inactive>
        Destinations : 0      Routes : 0
```

测试数据通信情况，当 R1 发送数据时，通过静态路由把数据包从 G0/0/0 接口或者 S1/0/0 接口正确地发送到 Internet 设备，由于 R1 发送数据的源地址为 202.100.1.1 或 12.1.1.1，而 Internet 设备默认有到达这两个地址的直连路由，所以数据可以正常通信。

```
<HW-R1>ping 8.8.8.8
   ping 8.8.8.8: 56    data bytes, press CTRL_C to break
      Reply from 8.8.8.8: bytes=56 Sequence=1 ttl=255 time=130 ms
```

```
Reply from 8.8.8.8: bytes=56 Sequence=2 ttl=255 time=20 ms
Reply from 8.8.8.8: bytes=56 Sequence=3 ttl=255 time=20 ms
Reply from 8.8.8.8: bytes=56 Sequence=4 ttl=255 time=20 ms
Reply from 8.8.8.8: bytes=56 Sequence=5 ttl=255 time=10 ms

--- 8.8.8.8 ping statistics ---
  5 packet(s) transmitted
  5 packet(s) received
  0.00% packet loss
  round-trip min/avg/max = 10/40/130 ms
```

注意，当出现 ECMP（等价负载多路径）时，大多数网络设备默认通过基于目的的转发结果，将源地址、目的 IP 地址等数据中的参数散列计算后转发到某一链路，散列结果不同，则会转发到不同链路。基于这种情况，某些关键业务要求流量必须从特定链路转发，这时就可以采用浮动静态路由来实现。

28.3.2 在网关设备上配置浮动静态路由

在图 28-4 中，G0/0/0 接口连接一条高速链路，而 S1/0/0 接口连接一条低速链路，此时需要把网络改造为浮动静态路由，使得到达 8.8.8.8 的流量从 G0/0/0 接口转发，当 G0/0/0 接口出现故障时，备份链路 S1/0/0 接口开始工作。浮动路由的本质是通过修改静态路由的优先级，使得优先级较高（数值较小）的链路出现在路由表中，其他链路作为备份链路并不装载到路由表中。

```
[HW-R1]dis cu | i ip route    //之前配置的静态路由
 ip route-static 8.8.8.8 255.255.255.255 Serial1/0/0    //串口的地址为 12.1.1.1/30
 ip route-static 8.8.8.8 255.255.255.255 GigabitEthernet0/0/0 202.100.1.2
[HW-R1]ip route-static 8.8.8.8 255.255.255.255 Serial1/0/0 preference 61    //修改低速链路的静态路由，优先级数值较大，则优先级较低，此链路将不会被装载到路由表中
Info: Succeeded in modifying route.
[HW-R1]display ip routing-table protocol static
Route Flags: R - relay, D - download to fib
------------------------------------------------------------------------------
Public routing table : Static
         Destinations : 1        Routes : 2        Configured Routes : 2

Static routing table status : <Active>
         Destinations : 1        Routes : 1

Destination/Mask    Proto    Pre    Cost    Flags    NextHop          Interface

     8.8.8.8/32     Static    60     0       D       202.100.1.2      GigabitEthernet0/0/0
//G0/0/0 接口的路由标识为<Active>，即该路由已被装载到路由表中

Static routing table status : <Inactive>
         Destinations : 1        Routes : 1
```

Destination/Mask	Proto	Pre	Cost	Flags	NextHop	Interface
8.8.8.8/32	Static	61	0		12.1.1.1	Serial1/0/0 //优先级为 61

的以 S1/0/0 接口作为出接口的路由被装载到备份的路由表中

测试数据通信情况，如下所示，此时数据只能通过 G0/0/0 接口转发。

```
[HW-R1]ping 8.8.8.8
  ping 8.8.8.8: 56   data bytes, press CTRL_C to break
    Reply from 8.8.8.8: bytes=56 Sequence=1 ttl=255 time=40 ms
    Reply from 8.8.8.8: bytes=56 Sequence=2 ttl=255 time=20 ms
    Reply from 8.8.8.8: bytes=56 Sequence=3 ttl=255 time=10 ms
    Reply from 8.8.8.8: bytes=56 Sequence=4 ttl=255 time=30 ms
    Reply from 8.8.8.8: bytes=56 Sequence=5 ttl=255 time=10 ms

  --- 8.8.8.8 ping statistics ---
    5 packet(s) transmitted
    5 packet(s) received
    0.00% packet loss
    round-trip min/avg/max = 10/22/40 ms
```

关闭 G0/0/0 接口来模拟网络链路故障，使得数据报文可以通过 S1/0/0 接口进行转发。

```
[HW-R1]interface g0/0/0
[HW-R1-GigabitEthernet0/0/0]shutdown
Dec 29 2018 17:13:56-08:00 HW-R1 %%01IFPDT/4/IF_STATE(l)[0]:Interface GigabitEthernet0/0/0 has turned into DOWN state.
Dec 29 2018 17:13:56-08:00 HW-R1 %%01IFNET/4/LINK_STATE(l)[1]:The line protocol IP on the interface GigabitEthernet0/0/0 has entered the DOWN state.
[HW-R1-GigabitEthernet0/0/0]display ip rout protocol static   //由于 G0/0/0 接口被关闭，此时依赖该
```
链路的静态路由会消失，另外一条备份路由会出现在路由表中

Route Flags: R - relay, D - download to fib

Public routing table : Static
         Destinations : 1   Routes : 1   Configured Routes : 2

Static routing table status : <Active>
         Destinations : 1   Routes : 1

Destination/Mask	Proto	Pre	Cost	Flags	NextHop	Interface
8.8.8.8/32	Static	61	0	D	12.1.1.1	Serial1/0/0 //到达 8.8.8.8

的路由此时显示优先级为 61，而非之前的出接口为 G0/0/0 接口的优先级为 60 的路由

Static routing table status : <Inactive>
         Destinations : 0   Routes : 0

```
[HW-R1-GigabitEthernet0/0/0]ping 8.8.8.8   //数据此时通过 S1/0/0 接口进行转发
  ping 8.8.8.8: 56   data bytes, press CTRL_C to break
    Reply from 8.8.8.8: bytes=56 Sequence=1 ttl=255 time=40 ms
```

```
Reply from 8.8.8.8: bytes=56 Sequence=2 ttl=255 time=20 ms
Reply from 8.8.8.8: bytes=56 Sequence=3 ttl=255 time=30 ms
Reply from 8.8.8.8: bytes=56 Sequence=4 ttl=255 time=20 ms
Reply from 8.8.8.8: bytes=56 Sequence=5 ttl=255 time=20 ms

--- 8.8.8.8 ping statistics ---
  5 packet(s) transmitted
  5 packet(s) received
  0.00% packet loss
  round-trip min/avg/max = 20/26/40 ms
```

28.3.3 在网关设备上配置浮动静态默认路由

在前面的案例中，演示了静态路由的实施方案，不得不说的是，静态路由往往用于网络的局部、临时调整，或者部署在企业网关设备上用于访问互联网。在图 28-4 中 R1 的位置配置静态路由并不合适。那么，如果用户需要访问互联网的其他地址，能通过到达 8.8.8.8 的路由去访问吗？显然是不行的，此时需要在企业网关设备上配置静态默认路由，即 0.0.0.0/0 的默认路由。

```
[HW-R1]int g0/0/0
[HW-R1-GigabitEthernet0/0/0]undo shutdown    //重新开启 G0/0/0 接口
[HW-R1]undo ip route-static 8.8.8.8 255.255.255.255 Serial1/0/0 preference 61
[HW-R1]undo ip route-static 8.8.8.8 255.255.255.255 GigabitEthernet0/0/0 202.100   //去除之前的静态明细路由配置
[HW-R1]ip route-static 0.0.0.0 0.0.0.0 GigabitEthernet 0/0/0    //配置浮动静态默认路由
```

注意，该方案存在一个非常严重的错误，在配置静态路由时，当选择使用加出接口而不带下一跳地址的配置方式时，所配置的网络被路由器作为直连网络处理，当目标地址为这些网络的第一个报文到达网络设备时，路由器会对这些目的 IP 地址进行 ARP 查找，但华为设备默认不开启代理 ARP，将导致数据转发失败。

```
[HW-R1]ip route-static 0.0.0.0 0.0.0.0 s1/0/0 preference 61    //配置浮动静态默认路由，即修改默认路由的优先级为 61，该路由不会被装载到路由表中
```

验证路由表：

```
[HW-R1]display ip routing-table protocol static
Route Flags: R - relay, D - download to fib
------------------------------------------------------------------------
Public routing table : Static
         Destinations : 1        Routes : 2        Configured Routes : 2

Static routing table status : <Active>
         Destinations : 1        Routes : 1

Destination/Mask    Proto    Pre  Cost        Flags  NextHop          Interface

        0.0.0.0/0   Static   60   0             D    202.100.1.1      GigabitEthernet0/0/0
Static routing table status : <Inactive>
         Destinations : 1        Routes : 1
```

```
Destination/Mask      Proto    Pre   Cost        Flags NextHop          Interface
        0.0.0.0/0     Static   61    0                 12.1.1.1         Serial1/0/0
```

```
<HW-R1>ping 8.8.8.8    //当通过 G0/0/0 接口的默认路由转发数据时，不能通信
   ping 8.8.8.8: 56   data bytes, press CTRL_C to break
     Request time out
     Request time out
     Request time out
     Request time out
     Request time out

   --- 8.8.8.8 ping statistics ---
     5 packet(s) transmitted
     0 packet(s) received
     100.00% packet loss
```

验证 ARP 表，由于华为设备默认不开启代理 ARP 功能，则设备上无法完成去往 8.8.8.8 的 MAC 地址封装，因而不能发送数据，如下所示：

```
<HW-R1>display arp
  IP ADDRESS         MAC ADDRESS          EXPIRE(M) TYPE            INTERFACE
VPN-INSTANCE
                                                    VLAN/CEVLAN PVC
------------------------------------------------------------------------
  202.100.1.1        00e0-fc39-67aa                I -             GE0/0/0
  10.1.11.1          00e0-fc39-67ab                I -             GE0/0/1
------------------------------------------------------------------------
  Total:2    Dynamic:0    Static:0    Interface:2
```

基于之前的分析，我们可以在 Internet 设备上开启代理 ARP 功能，注意，这是一个不恰当的解决方案。

```
[Internet-GigabitEthernet0/0/0]arp-proxy enable    //在下一跳设备开启代理 ARP 功能

<HW-R1>ping 8.8.8.8    //完成数据通信
   ping 8.8.8.8: 56   data bytes, press CTRL_C to break
     Reply from 8.8.8.8: bytes=56 Sequence=1 ttl=255 time=30 ms
     Reply from 8.8.8.8: bytes=56 Sequence=2 ttl=255 time=20 ms
     Reply from 8.8.8.8: bytes=56 Sequence=3 ttl=255 time=20 ms
     Reply from 8.8.8.8: bytes=56 Sequence=4 ttl=255 time=40 ms
     Reply from 8.8.8.8: bytes=56 Sequence=5 ttl=255 time=20 ms

   --- 8.8.8.8 ping statistics ---
     5 packet(s) transmitted
     5 packet(s) received
     0.00% packet loss
     round-trip min/avg/max = 20/26/40 ms
```

验证 R1 的 ARP 表：

```
<HW-R1>display arp    //可以看到到达 8.8.8.8 这个远程网络封装的 MAC 地址 00e0-fc6e-5b46 实际
```

上是 Internet 设备的 G0/0/0 接口的 MAC 地址，虽然此时可以通信，但每当出现一个到达某网络的地址时，就进行一次 ARP 查询，也会在设备上形成一个 ARP 表项，这无疑大大加重了设备的负载，所以必须杜绝这种操作。

```
IP ADDRESS        MAC ADDRESS        EXPIRE(M) TYPE         INTERFACE
VPN-INSTANCE
                                     VLAN/CEVLAN PVC
------------------------------------------------------------------------
202.100.1.1       00e0-fc39-67aa               I -          GE0/0/0
8.8.8.8           00e0-fc6e-5b46   20          D-0          GE0/0/0
10.1.11.1         00e0-fc39-67ab               I -          GE0/0/1
------------------------------------------------------------------------
Total:3           Dynamic:1        Static:0    Interface:2
```

正确的实施方案如下：

```
[HW-R1]undo ip route-static 0.0.0.0 0.0.0.0 GigabitEthernet 0/0/0   //去除错误的配置（手动去除），否则可能出现多条路由
[HW-R1]ip route-static 0.0.0.0 0.0.0.0 GigabitEthernet 0/0/0 202.100.1.2   //配置正确的静态默认路由
[HW-R1]dis ip rou pro static
Route Flags: R - relay, D - download to fib
------------------------------------------------------------------------
Public routing table : Static
         Destinations : 1      Routes : 2       Configured Routes : 2

Static routing table status : <Active>
         Destinations : 1      Routes : 1

Destination/Mask    Proto    Pre   Cost    Flags NextHop         Interface

     0.0.0.0/0      Static   60    0        D    202.100.1.2     GigabitEthernet0/0/0

Static routing table status : <Inactive>
         Destinations : 1      Routes : 1

Destination/Mask    Proto    Pre   Cost    Flags NextHop         Interface

     0.0.0.0/0      Static   61    0             12.1.1.1        Serial1/0/0

<HW-R1>reset arp all   //重置 ARP 表
Warning: This operation will reset all static and dynamic ARP entries, and clear the configurations of all static ARP, continue?[Y/N]:y

<HW-R1>ping 8.8.8.8    //测试数据通信情况，一切正常
  ping 8.8.8.8: 56   data bytes, press CTRL_C to break
    Reply from 8.8.8.8: bytes=56 Sequence=1 ttl=255 time=40 ms
    Reply from 8.8.8.8: bytes=56 Sequence=2 ttl=255 time=20 ms
    Reply from 8.8.8.8: bytes=56 Sequence=3 ttl=255 time=20 ms
    Reply from 8.8.8.8: bytes=56 Sequence=4 ttl=255 time=30 ms
```

```
    --- 8.8.8.8 ping statistics ---
    4 packet(s) transmitted
    4 packet(s) received
    0.00% packet loss
    round-trip min/avg/max = 20/27/40 ms
<HW-R1>dis arp    //IP 地址 8.8.8.8 已经没有对应的 MAC 地址映射,而只有正确的下一跳 202.100.1.2
的 MAC 地址解析
    IP ADDRESS        MAC ADDRESS        EXPIRE(M) TYPE           INTERFACE
VPN-INSTANCE
                                         VLAN/CEVLAN PVC
    ------------------------------------------------------------------------
    202.100.1.1       00e0-fc39-67aa               I -            GE0/0/0
    202.100.1.2       00e0-fc6e-5b46     20        D-0            GE0/0/0
    10.1.11.1         00e0-fc39-67ab               I -            GE0/0/1
    ------------------------------------------------------------------------
    Total:3           Dynamic:1          Static:0  Interface:2
```

本案例实施完毕。

案例 29　在思科设备上配置静态路由

29.1　静态路由配置案例及拓扑

在企业网络的网关设备上配置静态路由、浮动静态路由和默认路由，保证数据可以转发到互联网。如图 29-1 所示，使用思科设备在企业网关 R1 上配置静态路由、浮动静态路由和默认路由。

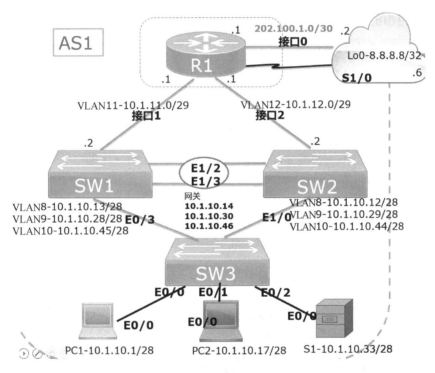

图 29-1　使用思科设备在企业网关 R1 上配置静态路由、浮动静态路由和默认路由

29.2　静态路由配置要点

配置静态路由的要点如下：
① 在保证接口工作的前提下配置静态明细路由。
② 配置静态浮动路由，理解优先级的作用。
③ 配置默认路由，用以访问互联网，理解下一跳和出接口的异同。

29.3 静态路由配置步骤详解

29.3.1 在网关设备上配置静态明细路由

在企业网的网关设备即 R1 上配置静态路由，访问互联网地址 8.8.8.8。本节内容主要是为了帮助大家理解静态路由的作用，正确地完成配置工作的相关内容请参考后续章节。

C-R1(config)#ip route 8.8.8.8 255.255.255.255 e0/0 202.100.1.2   //手动配置到达 8.8.8.8/32 网络的出接口为 E0/0，下一跳地址为 202.100.1.2，这意味着到达地址 8.8.8.8 的数据从 E0/0 转发出去，转到下一个转发设备，这个设备的地址为 202.100.1.2，多点接入网络推荐使用出接口+下一跳的方式实现

C-R1(config)#ip route 8.8.8.8 255.255.255.255 s1/0   //配置到达 8.8.8.8/32 网络的出接口为 S1/0，由于这是一个点到点的网络，所以不需要配置下一跳地址，因为点到点接口只能把数据转给唯一的下一跳地址，而不像多点设备那样，可能下一跳地址并不是正确的转发设备

验证静态路由的配置结果：

```
C-R1#show ip route
Codes: L - local, C - connected, S - static, R - RIP, M - mobile, B - BGP
       D - EIGRP, EX - EIGRP external, O - OSPF, IA - OSPF inter area
       N1 - OSPF NSSA external type 1, N2 - OSPF NSSA external type 2
       E1 - OSPF external type 1, E2 - OSPF external type 2
       i - IS-IS, su - IS-IS summary, L1 - IS-IS level-1, L2 - IS-IS level-2
       ia - IS-IS inter area, * - candidate default, U - per-user static route
       o - ODR, P - periodic downloaded static route, H - NHRP, l - LISP
       a - application route
       + - replicated route, % - next hop override

Gateway of last resort is not set

      8.0.0.0/32 is subnetted, 1 subnets
S        8.8.8.8 [1/0] via 202.100.1.2, Ethernet0/0    //验证到 8.8.8.8/32 网络的路由的下一跳和出接
```
口配置结果，静态路由已经被装载到路由表中

```
                is directly connected, Serial1/0    //验证到达 8.8.8.8/32 网络直接出接口的路由配
```
置结果，即达到 8.8.8.8 网络的数据直接从 S1/0 发送出去。由于到达 8.8.8.8/32 网络的静态路由有两条，所以会在路由表中出现两条路由，实现路由负载均衡

```
      10.0.0.0/8 is variably subnetted, 2 subnets, 2 masks
C        10.1.11.0/29 is directly connected, Ethernet0/1
L        10.1.11.1/32 is directly connected, Ethernet0/1
      12.0.0.0/8 is variably subnetted, 2 subnets, 2 masks
C        12.1.1.0/30 is directly connected, Serial1/0
L        12.1.1.1/32 is directly connected, Serial1/0
      202.100.1.0/24 is variably subnetted, 2 subnets, 2 masks
C        202.100.1.0/30 is directly connected, Ethernet0/0
L        202.100.1.1/32 is directly connected, Ethernet0/0
```
//只有在设备具备路由表之后，路由器才可以依据这些路由表进行决策，进而把数据从对应的出接口转发出去，到达下一跳设备

C-R1#ping 8.8.8.8   //数据可以正常发送到目的地 8.8.8.8
Type escape sequence to abort.

```
Sending 5, 100-byte ICMP Echos to 8.8.8.8, timeout is 2 seconds:
!!!!!
Success rate is 100 percent (5/5), round-trip min/avg/max = 1/1/2 ms
C-R1#show ip route 8.8.8.8
Routing entry for 8.8.8.8/32
  Known via "static", distance 1, metric 0 (connected)
  Routing Descriptor Blocks:
  * 202.100.1.2, via Ethernet0/0    //在思科设备上，如果出现路由负载均衡，那么将显示一个"*"，
这意味着设备将会使用这条路由转发数据
      Route metric is 0, traffic share count is 1
    directly connected, via Serial1/0
      Route metric is 0, traffic share count is 1
```

路由负载均衡的优势之一在于，当一条路由失效之后，其他路由依旧可以转发数据。如下所示，当我们关闭以太网接口时，数据依旧可以通过 S1/0 转发到目的地。

```
C-R1(config)#int e0/0
C-R1(config-if)#shutdown
C-R1#show ip route   8.8.8.8    //可以看到到达 8.8.8.8 的另外一条路由依旧存在
Routing entry for 8.8.8.8/32
  Known via "static", distance 1, metric 0 (connected)
  Routing Descriptor Blocks:
  * directly connected, via Serial1/0
      Route metric is 0, traffic share count is 1
C-R1#ping 8.8.8.8
Type escape sequence to abort.
Sending 5, 100-byte ICMP Echos to 8.8.8.8, timeout is 2 seconds:
!!!!!
Success rate is 100 percent (5/5), round-trip min/avg/max = 9/9/10 ms
```

实验完毕，请自行开启以太网接口。

29.3.2 在网关设备上配置浮动静态路由

本节我们依旧配置静态方式的明细路由，但为了更好地理解管理距离，将修改管理距离的优先级，使得网络转发存在更明确的路径。

```
C-R1(config)#ip route 8.8.8.8 255.255.255.255 s1/0 10    //修改到达 8.8.8.8/32 路由的管理距离，将默认
值 1 修改为 10，那么另外一条出接口为 E0/0 的路由将会"独自"出现在路由表中（默认值还是 1）
C-R1#show ip route
Codes: L - local, C - connected, S - static, R - RIP, M - mobile, B - BGP
       D - EIGRP, EX - EIGRP external, O - OSPF, IA - OSPF inter area
       N1 - OSPF NSSA external type 1, N2 - OSPF NSSA external type 2
       E1 - OSPF external type 1, E2 - OSPF external type 2
       i - IS-IS, su - IS-IS summary, L1 - IS-IS level-1, L2 - IS-IS level-2
       ia - IS-IS inter area, * - candidate default, U - per-user static route
       o - ODR, P - periodic downloaded static route, H - NHRP, l - LISP
       a - application route
       + - replicated route, % - next hop override
```

```
Gateway of last resort is not set

        8.0.0.0/32 is subnetted, 1 subnets
S        8.8.8.8 is directly connected, Ethernet0/0   //到达 8.8.8.8/32 的路由仅剩下以 E0/0 为出接口
的路由，而管理距离被修改为较大值的路由只有在路由表中的相关路由消失之后才被加载
        10.0.0.0/8 is variably subnetted, 2 subnets, 2 masks
C        10.1.11.0/29 is directly connected, Ethernet0/1
L        10.1.11.1/32 is directly connected, Ethernet0/1
        12.0.0.0/8 is variably subnetted, 2 subnets, 2 masks
C        12.1.1.0/30 is directly connected, Serial1/0
L        12.1.1.1/32 is directly connected, Serial1/0   //该地址是串行接口的地址
        202.100.1.0/24 is variably subnetted, 2 subnets, 2 masks
C        202.100.1.0/30 is directly connected, Ethernet0/0
 --More--
*Jan  2 07:10:47.127: %SYS-5-CONFIG_I: Configured from console by console
L        202.100.1.1/32 is directly connected, Ethernet0/0
```

验证当主路由消失时，备份路由被装载的效果。

```
C-R1(config)#interface e0/0
C-R1(config-if)#shutdown   //关闭 R1 的 E0/0 接口模拟主路由失效的场景（注意不要关闭 Internet
设备的 E0/0 接口，思科模拟器的一个常见 bug 就是当对端的以太网接口关闭后，本端的以太网接口并不关
闭，如果是真实设备，对端以太网接口关闭，本端以太网接口也会关闭）
C-R1(config-if)#
*Jan  2 07:17:48.065: %LINK-5-CHANGED: Interface Ethernet0/0, changed state to administratively
down
*Jan  2 07:17:48.676: %SYS-5-CONFIG_I: Configured from console by console
*Jan  2 07:17:49.070: %LINEPROTO-5-UPDOWN: Line protocol on Interface Ethernet0/0, changed
state to down
C-R1#show ip route static
Codes: L - local, C - connected, S - static, R - RIP, M - mobile, B - BGP
       D - EIGRP, EX - EIGRP external, O - OSPF, IA - OSPF inter area
       N1 - OSPF NSSA external type 1, N2 - OSPF NSSA external type 2
       E1 - OSPF external type 1, E2 - OSPF external type 2
       i - IS-IS, su - IS-IS summary, L1 - IS-IS level-1, L2 - IS-IS level-2
       ia - IS-IS inter area, * - candidate default, U - per-user static route
       o - ODR, P - periodic downloaded static route, H - NHRP, l - LISP
       a - application route
       + - replicated route, % - next hop override


Gateway of last resort is not set

        8.0.0.0/32 is subnetted, 1 subnets
S        8.8.8.8 is directly connected, Serial1/0
C-R1#ping 8.8.8.8   //数据依旧可以通过备份链路 S1/0 转发，验证完毕，请读者自行开启以太网接口
Type escape sequence to abort.
Sending 5, 100-byte ICMP Echos to 8.8.8.8, timeout is 2 seconds:
!!!!!
Success rate is 100 percent (5/5), round-trip min/avg/max = 9/9/11 ms
```

浮动静态路由的本质是修改管理距离，对同一条路由，管理距离小的会被装载到路由表中，管理距离大的以备份路由形式出现。后面我们还会学习动态路由协议如何通过修改管理距离来影响路由表的状态。

29.3.3 在网关设备上配置浮动静态默认路由

在上一个实验中通过静态路由，设备可以访问地址 8.8.8.8，但如何利用同样的方法访问其他互联网地址呢？这些互联网地址何止成千上万，所以明细路由并不能完全满足客户访问互联网的要求，此时需要配置默认路由。我们在企业网关设备 R1 上配置默认路由。

  C-R1(config)#ip route 0.0.0.0 0.0.0.0 s1/0 10　　//到达任意网络（默认路由）的出接口为 S1/0，备用链路由的管理距离较大，并不能被装载到路由表中
  C-R1(config)#ip route 0.0.0.0 0.0.0.0 e0/0 202.100.1.2　　//到达任意网络（默认路由）的出接口为 E0/0，下一跳地址为 202.100.1.2（即运营商的一个公网地址）
  C-R1#show ip route static　　//验证路由表
  Codes: L - local, C - connected, S - static, R - RIP, M - mobile, B - BGP
     D - EIGRP, EX - EIGRP external, O - OSPF, IA - OSPF inter area
     N1 - OSPF NSSA external type 1, N2 - OSPF NSSA external type 2
     E1 - OSPF external type 1, E2 - OSPF external type 2
     i - IS-IS, su - IS-IS summary, L1 - IS-IS level-1, L2 - IS-IS level-2
     ia - IS-IS inter area, * - candidate default, U - per-user static route
     o - ODR, P - periodic downloaded static route, H - NHRP, l - LISP
     a - application route
     + - replicated route, % - next hop override

  Gateway of last resort is 202.100.1.2 to network 0.0.0.0

  S* 0.0.0.0/0 is directly connected, Ethernet0/0　　//默认路由出接口为 E0/0
     8.0.0.0/32 is subnetted, 1 subnets
  S   8.8.8.8 is directly connected, Ethernet0/0　　//明细路由，即到达 8.8.8.8 的路由

依据最长匹配原则，如果数据包中目的 IP 地址为 8.8.8.8，则采用下面的路由；如果目的 IP 地址为其他网络（比如 7.7.7.7 等），则使用上面的默认路由。

以上是默认路由的推荐配置，接下来验证在一条静态路由中仅配置出接口的情况：

  C-R1(config)#no ip route 0.0.0.0 0.0.0.0 Ethernet0/0 202.100.1.2　　//去除掉之前配置的推荐实施的默认路由
  C-R1(config)#ip route 0.0.0.0 0.0.0.0 Ethernet0/0　　//该配置方案仅配置了出接口，注意思科的 IOS 会有一条日志出现（不得不说，IT 人员最基本的技能就是看懂日志），该日志表明"在配置一条默认路由时，如果不是一个点到点的接口，而仅配置了出接口，将大大影响性能"
  %Default route without gateway, if not a point-to-point interface, may impact performance

思科设备默认开启代理 ARP 功能，所以此时可以实现数据通信，如下所示：

  C-R1#ping 7.7.7.7　　//发送到达 IP 地址为 7.7.7.7 的数据，R1 会以默认路由作为决策依据，而默认路由仅配置了出接口，会进行对应的地址 7.7.7.7 的 ARP 查询，然后将数据从 E0/0 接口发送出去，此时由于 Internet 的 E0/0 接口开启了代理 ARP 功能，它会以 Internet 设备 E0/0 接口的 MAC 地址作为 ARP 响应（前提是 Internet 设备有到达地址 7.7.7.7 的路由）
  Type escape sequence to abort.
  Sending 5, 100-byte ICMP Echos to 7.7.7.7, timeout is 2 seconds:

....
Success rate is 0 percent (0/4)
C-R1#sh arp
Protocol	Address	Age (min)	Hardware Addr	Type	Interface
Internet	7.7.7.7	0	Incomplete	ARPA	//此时由于 Internet 设备没有

到达 IP 地址 7.7.7.7 的路由，所以不会做出响应 |
Internet	10.1.11.1	-	aabb.cc00.1010	ARPA	Ethernet0/1
Internet	202.100.1.1	-	aabb.cc00.1000	ARPA	Ethernet0/0
Internet	202.100.1.2	0	aabb.cc00.2000	ARPA	Ethernet0/0

R2-internet(config)#interface lo1
R2-internet(config-if)#ip address 7.7.7.7 255.255.255.255
C-R1#ping 7.7.7.7
Type escape sequence to abort.
Sending 5, 100-byte ICMP Echos to 7.7.7.7, timeout is 2 seconds:
!!!!!
Success rate is 100 percent (5/5), round-trip min/avg/max = 1/1/1 ms
C-R1#sh arp
Protocol	Address	Age (min)	Hardware Addr	Type	Interface
Internet	7.7.7.7	0	aabb.cc00.2000	ARPA	Ethernet0/0　//到达 7.7.7.7 的

ARP 表项封装了 Internet 设备 E0/0 接口的 MAC 地址 |
Internet	10.1.11.1	-	aabb.cc00.1010	ARPA	Ethernet0/1
Internet	202.100.1.1	-	aabb.cc00.1000	ARPA	Ethernet0/0
Internet	202.100.1.2	0	aabb.cc00.2000	ARPA	Ethernet0/0

验证完毕，请自行去掉该配置，改回推荐配置。可以通过以下操作查看配置验证是否成功：

C-R1(config)#do sh run | s ip route
ip route 0.0.0.0 0.0.0.0 Ethernet0/0 202.100.1.2
ip route 0.0.0.0 0.0.0.0 Serial1/0 10
ip route 8.8.8.8 255.255.255.255 Ethernet0/0 202.100.1.2
ip route 8.8.8.8 255.255.255.255 Serial1/0 10

为方便后续实验，请保存默认路由的配置。

本案例实施完毕。

案例 30　在华为设备上配置 OSPF

OSPF 路由协议简书

IETF（Internet Engineering Task Force，互联网工程任务组）提出了基于 SPF 算法的链路状态 OSPF（Open Shortest Path First，开放最短路径优先）路由协议。通过在大型网络中部署 OSPF 协议，弥补了 RIP 协议的诸多不足。作为一种公有标准，OSPF 协议是企业网应用最普遍的协议，它的 IP 协议号是 89，其目的是创建 IP 路由表，使得路由器完成决策后进一步转发数据。OSPF 协议作为链路状态路由协议，不直接传递各路由器的路由表，而传递链路状态信息，各路由器基于链路状态信息独立计算路由。所有路由器各自维护一个链路状态数据库。邻居路由器间先同步链路状态数据库，再各自基于 SPF（Shortest Path First，最短路径优先）算法计算最优路由，从而提高收敛速度。在度量方式上，OSPF 协议（简称 OSPF）将链路带宽作为选路时的参考依据。"累计带宽"是一种比"累计跳数"更科学的计算方式。

OSPF 的路由计算过程可以简单描述如下。

- 路由器之间发现并建立邻居关系。
- 每台路由器产生并向邻居泛洪链路状态信息，同时收集来自其他路由器的链路状态信息，完成 LSDB（Link State Data Base，链路状态数据库）的同步。
- 每台路由器基于 LSDB 通过 SPF 算法，计算得到一棵以自己为根的 SPT（Shortest Path Tree，最短路径树），再以 SPT 为基础计算去往各邻居连接网络的最优路由，并形成路由表。

OSPF 邻居、数据库和路由计算过程如图 30-1 所示。

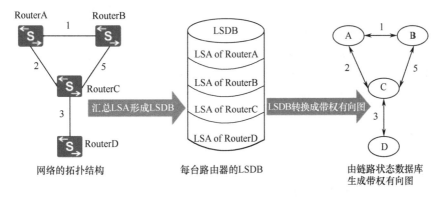

图 30-1　OSPF 邻居、数据库和路由计算过程

（1）区域（Area）

OSPF 支持将一组网段组合在一起，这样的一个组合称为一个区域，即区域是一组网段

的集合。划分区域可以缩小 LSDB 的规模，减少网络流量。

区域内的详细拓扑信息不向其他区域发送，区域间传递的是抽象的路由信息，而不是详细的描述拓扑结构的链路状态信息。每个区域都有自己的 LSDB，不同区域的 LSDB 是不同的。路由器会为每一个自己所连接到的区域维护一个单独的 LSDB。由于详细链路状态信息不会被发布到区域以外，因此 LSDB 的规模大大缩小了。

Area 0 为骨干区域。骨干区域负责在非骨干区域之间发布由区域边界路由器汇总的路由信息（并非详细的链路状态信息），为了避免出现区域间路由环路，非骨干区域之间不允许直接相互发布区域间路由信息。因此，所有区域边界路由器都至少有一个接口属于 Area 0，即每个区域都必须连接到骨干区域。

（2）OSPF 支持的网络类型

OSPF 之所以可以在企业网中广泛应用，是因为它支持多种多样的网络类型，常见的网络类型有以下 4 种，分别是广播型网络、NBMA 网络、点到多点网络和点到点网络。

- 广播（Broadcast）型网络：当链路层协议是 Ethernet 时，OSPF 默认网络类型是 Broadcast。在该类型的网络中，通常以组播形式发送 Hello 报文、LSU 报文和 LSAck 报文。其中，224.0.0.5 为 OSPF 设备的预留 IP 组播地址；224.0.0.6 为 OSPF DR/BDR 的预留 IP 组播地址。在此类型网络中，以单播形式发送 DD 报文和 LSR 报文，同时会选择 DR（指定路由器）和 BDR（备份指定路由器）。
- NBMA（Non-Broadcast Multi-Access，非广播-多路访问）型网络：当链路层协议是帧中继或 X.25 协议时，默认情况下，OSPF 认为网络类型是 NBMA 类型。在该类型的网络中，以单播形式发送协议报文（Hello 报文、DD 报文、LSR 报文、LSU 报文、LSAck 报文），同时会选择 DR（指定路由器）和 BDR（备份指定路由器）。
- P2MP（Point-to-Multipoint，点到多点）型网络：由于在链路层协议中没有 P2MP 概念，所以点到多点网络必须是由其他的网络类型强制更改的。常用做法是将非全连通的 NBMA 网络改为点到多点的网络。在该类型的网络中，以组播形式（224.0.0.5）发送 Hello 报文，以单播形式发送其他协议报文（DD 报文、LSR 报文、LSU 报文、LSAck 报文）。
- P2P（Point-to-Point，点到点）型网络：当链路层协议是 PPP 链路、HDLC 链路或 LAPB 链路协议时，在默认情况下，OSPF 认为网络类型是 P2P 类型。
- 虚链路也被认为是一种网络类型（本书不涉及）。

思科设备还支持特有的 P2MP 非广播的网络类型。

（3）影响 OSPF 邻居关系建立的因素

- MA 网段的掩码需要一致。
- 相邻设备的 RID 不能相同。
- 链路两侧 Area ID 必须一致。
- Hello Time 和 Dead Time 都要一致。
- 接口的 MTU 一致（华为设备默认情况下不检测 DBD 报文中的 MTU，思科设备默认检测）。

- 认证类型和认证数据需要一致。
- 区域类型需要一致，特殊区域（NSSA 和 STUB）不能和普通区域建立邻居。

注意：链路两侧不一致的 OSPF 进程不影响邻居关系的建立。

（4）DR 和 BDR 的作用

① 减少邻接关系的数量，从而减少链路状态信息及路由信息的交换次数，这样可以节省带宽，减少路由器硬件的负担。一个既不是 DR 也不是 BDR 的路由器只与 DR 和 BDR 形成邻接关系并交换链路状态信息和路由信息，这样就大大减少了大型广播型网络和 NBMA 网络中的邻接关系数量。

② 在描述拓扑的 LSDB 中，一个 NBMA 网段或广播型网段是由单独一条 LSA 来描述的，这条 LSA 由该网段上的 DR 产生。

（5）选举和切换

- DR 和 BDR 由 OSPF 的 Hello 协议选举，选举是根据接口上的路由器优先级（Router Priority）进行的。
- 如果 Router Priority 被设置为 0，那么该路由器将不允许被选举为 DR 或 BDR。
- Router Priority 的值越大优先级越高。如果 Router Priority 的值相同，Router ID 大者优先级高。
- 为了维护网络上邻接关系的稳定性，如果网络中已经存在 DR 和 BDR，则新添加进该网段的路由器不会成为 DR 和 BDR，不管该路由器的 Router Priority 值是否最大。

如果当前 DR 发生故障，当前 BDR 自动成为新的 DR，网络中将重新选举 BDR；如果当前 BDR 发生故障，则 DR 不变，重新选举 BDR。这种选举机制的目的是保持邻接关系的稳定性，减小拓扑结构的改变对邻接关系的影响。

DR 是基于广播型接口的，这意味着如果有多个广播型接口则可能存在多个 DR，它不是一个基于设备的概念，这意味着并不是一个设备只有一个 DR。

30.1　在企业内部网络上配置 OSPF

在企业网内部配置多区域的 OSPF 以保证企业内部网络通信，同时为后续企业网络接入互联网做好准备。

图 30-2 所示为在企业园区 AS1 内配置多区域 OSPF 的拓扑，其中 R1 的环回接口 0 位于 OSPF 区域 0（骨干区域），R1 的其他内网接口（接口 1 和接口 2），以及 SW1 和 SW2 上所有的 VLANIF 运行在 OSPF 的区域 1。请确保所有设备可以建立基本的邻居关系，之后会发现设备能够自动学习路由。

30.2　配置 OSPF 的要点

① 保证设备互连接口工作且直连网络通信正常。
② 创建 OSPF 进程并配置 RID。

③ 确保接口运行 OSPF 并配置正确的区域。
④ 验证和查验 OSPF 状态，完成数据通信。

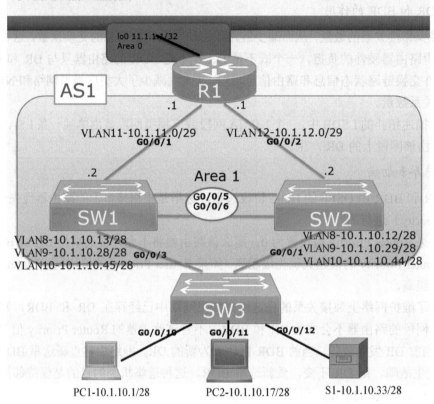

图 30-2 在企业园区 AS1 内配置多区域 OSPF 的拓扑

30.3 配置 OSPF 的步骤详解

30.3.1 在重要的网络设备之间配置 OSPF

请测试直连网络的连通性，通常在保证直连通信的情况下完成 OSPF 配置。另外，在本例中会用到一个新的逻辑接口——环回（Loopback）接口，其作用如图 30-3 所示（具体内容请关注笔者的其他书籍和课程）。

测试直连网络的通信情况：

```
<R1>ping 10.1.11.2
  ping 10.1.11.2: 56    data bytes, press CTRL_C to break
    Reply from 10.1.11.2: bytes=56 Sequence=1 ttl=255 time=30 ms
    Reply from 10.1.11.2: bytes=56 Sequence=2 ttl=255 time=40 ms
    Reply from 10.1.11.2: bytes=56 Sequence=3 ttl=255 time=10 ms
    Reply from 10.1.11.2: bytes=56 Sequence=4 ttl=255 time=30 ms
    Reply from 10.1.11.2: bytes=56 Sequence=5 ttl=255 time=20 ms
```

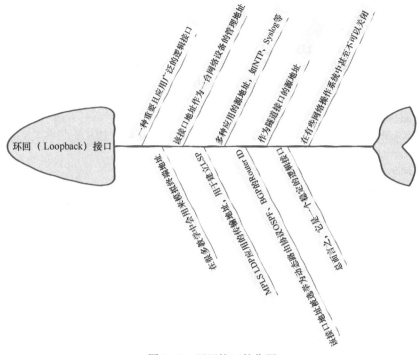

图 30-3　环回接口的作用

```
    --- 10.1.11.2 ping statistics ---
    5 packet(s) transmitted
    5 packet(s) received
    0.00% packet loss
    round-trip min/avg/max = 10/26/40 ms

<R1>ping 10.1.12.2
  ping 10.1.12.2: 56   data bytes, press CTRL_C to break
    Reply from 10.1.12.2: bytes=56 Sequence=1 ttl=255 time=20 ms
    Reply from 10.1.12.2: bytes=56 Sequence=2 ttl=255 time=20 ms
    Reply from 10.1.12.2: bytes=56 Sequence=3 ttl=255 time=20 ms
    Reply from 10.1.12.2: bytes=56 Sequence=4 ttl=255 time=20 ms
    Reply from 10.1.12.2: bytes=56 Sequence=5 ttl=255 time=20 ms

    --- 10.1.12.2 ping statistics ---
    5 packet(s) transmitted
    5 packet(s) received
    0.00% packet loss
    round-trip min/avg/max = 20/20/20 ms

R1:
  ospf 1 router-id 1.1.1.1   //启动 OSPF 进程，推荐手动配置 OSPF 的路由器 ID，它是一个 IPv4 格式
的标识，此处配置为 1.1.1.1
    area 0.0.0.0    //创建区域 0，即骨干区域
    area 0.0.0.1    //创建区域 1，即一个常规区域，在本例以及后续的实施中，推荐在接口上使能 OSPF
```

协议，但是必须创建区域，在接口上实施才会生效，否则在华为设备上不会实施成功
```
        interface LoopBack0
         ip address 11.1.1.1 255.255.255.255
         ospf enable 1 area 0.0.0.0    //环回接口配置在 OSPF 区域 0，即骨干区域，同时该接口 IP 地址所在
网络被通告到 OSPF 区域，以便其他设备学习到这些路由（1.1.1.1/32）
        interface GigabitEthernet0/0/1
         ip address 10.1.11.1 255.255.255.248
         ospf enable 1 area 0.0.0.1    //设备互连接口配置在 OSPF 区域 1 中
        interface GigabitEthernet0/0/2
         ip address 10.1.12.1 255.255.255.248
         ospf enable 1 area 0.0.0.1
        SW1：
        ospf 1 router-id 11.11.11.11    //启动 OSPF 进程 1 并手动设置 Router ID 为 11.11.11.11
         area 0.0.0.1    //创建区域 1，之后在接口配置对应的区域 1，否则接口配置无效
        interface Vlanif11
         ip address 10.1.11.2 255.255.255.248
         ospf enable 1 area 0.0.0.1    //设备互连接口配置在 OSPF 区域 1 中，注意设备互连接口需要配置在
一致的区域 ID
        interface Vlanif8
         ip address 10.1.10.13 255.255.255.240
         vrrp vrid 8 virtual-ip 10.1.10.14
         vrrp vrid 8 priority 110
         ospf enable 1 area 0.0.0.1    //业务接口配置在 OSPF 区域 1 中
        interface Vlanif9
         ip address 10.1.10.28 255.255.255.240
         vrrp vrid 9 virtual-ip 10.1.10.30
         vrrp vrid 9 priority 110
         ospf enable 1 area 0.0.0.1
        interface Vlanif10
         ip address 10.1.10.45 255.255.255.240
         vrrp vrid 10 virtual-ip 10.1.10.46
         vrrp vrid 10 priority 110
         ospf enable 1 area 0.0.0.1
        #
        SW2：
        ospf 1 router-id 12.12.12.12
         area 0.0.0.1
        interface Vlanif12
         ip address 10.1.12.2 255.255.255.248
         ospf enable 1 area 0.0.0.1
        interface Vlanif8
         ip address 10.1.10.12 255.255.255.240
         vrrp vrid 8 virtual-ip 10.1.10.14
         ospf enable 1 area 0.0.0.1
        interface Vlanif9
         ip address 10.1.10.29 255.255.255.240
         vrrp vrid 9 virtual-ip 10.1.10.30
```

```
  ospf enable 1 area 0.0.0.1

interface Vlanif10
  ip address 10.1.10.44 255.255.255.240
  vrrp vrid 10 virtual-ip 10.1.10.46
  ospf enable 1 area 0.0.0.1
```

在通常情况下，OSPF 建立邻居关系后，会同步更新 OSPF 的链路状态数据库，之后会自动计算 OSPF 方式的路由。

验证 OSPF 运行的接口，请注意对应区域的配置要正确。

```
<R1>display ospf int

          OSPF Process 1 with Router ID 1.1.1.1
                Interfaces

Area: 0.0.0.0           (MPLS TE not enabled)
 IP Address    Type          State       Cost    Pri      DR         BDR
 11.1.1.1      P2P           P-2-P       0       1        0.0.0.0    0.0.0.0

Area: 0.0.0.1           (MPLS TE not enabled)
 IP Address    Type          State       Cost    Pri      DR         BDR
 10.1.11.1     Broadcast     P-2-P       1       1        0.0.0.0    0.0.0.0
 10.1.12.1     Broadcast     P-2-P       1       1        0.0.0.0    0.0.0.0
<R1>display ospf peer brief    //查看 OSPF 的邻居关系，请确保邻居关系正确，可以看到 R1 和其他
```
设备在区域 1 中构建了邻居关系

```
           OSPF Process 1 with Router ID 1.1.1.1
                 Peer Statistic Information
 ----------------------------------------------------------------
 Area Id      Interface                Neighbor id       State
 0.0.0.1      GigabitEthernet0/0/1     11.11.11.11       Full
 0.0.0.1      GigabitEthernet0/0/2     12.12.12.12       Full
 ----------------------------------------------------------------
<R1>display ip routing-table protocol ospf
Route Flags: R - relay, D - download to fib
 ----------------------------------------------------------------
Public routing table : OSPF
        Destinations : 6      Routes : 9

OSPF routing table status : <Active>
        Destinations : 6      Routes : 9

Destination/Mask      Proto    Pre   Cost    Flags    NextHop      Interface

     10.1.10.0/28     OSPF     10    2       D        10.1.12.2    GigabitEthernet0/0/2
                     OSPF     10    2       D        10.1.11.2    GigabitEthernet0/0/1
    10.1.10.14/32    OSPF     10    2       D        10.1.11.2    GigabitEthernet0/0/1
```

10.1.10.16/28	OSPF	10	2	D	10.1.12.2	GigabitEthernet0/0/2
	OSPF	10	2	D	10.1.11.2	GigabitEthernet0/0/1
10.1.10.30/32	OSPF	10	2	D	10.1.11.2	GigabitEthernet0/0/1
10.1.10.32/28	OSPF	10	2	D	10.1.12.2	GigabitEthernet0/0/2
	OSPF	10	2	D	10.1.11.2	GigabitEthernet0/0/1
10.1.10.46/32	OSPF	10	2	D	10.1.11.2	GigabitEthernet0/0/1

OSPF routing table status : <Inactive>
    Destinations : 0   Routes : 0

查看 SW1 和 SW2 的状态：

```
<HW-SW1>display ospf interface

   OSPF Process 1 with Router ID 11.11.11.11
           Interfaces

Area: 0.0.0.1
IP Address      Type          State    Cost    Pri    DR         BDR
10.1.10.13      <HW-SW2>display ospf interface

   OSPF Process 1 with Router ID 12.12.12.12
           Interfaces

Area: 0.0.0.1
IP Address      Type          State    Cost    Pri    DR         BDR
10.1.10.12      Broadcast     P-2-P    1       1      0.0.0.0    0.0.0.0
10.1.10.29      Broadcast     P-2-P    1       1      0.0.0.0    0.0.0.0
10.1.10.44      Broadcast     P-2-P    1       1      0.0.0.0    0.0.0.0
10.1.12.2       Broadcast     P-2-P    1       1      0.0.0.0    0.0.0.0

<HW-SW2>dis
<HW-SW2>display ip rou
<HW-SW2>display ip routing-table pro
<HW-SW2>display ospf peer bri
<HW-SW2>display ospf peer brief

    OSPF Process 1 with Router ID 12.12.12.12
            Peer Statistic Information
 ----------------------------------------------------------------
 Area Id       Interface        Neighbor id       State
 0.0.0.1       Vlanif8          11.11.11.11       Full
 0.0.0.1       Vlanif9          11.11.11.11       Full
 0.0.0.1       Vlanif10         11.11.11.11       Full
 0.0.0.1       Vlanif12         1.1.1.1           Full
 ----------------------------------------------------------------

<HW-SW1>display ospf peer brief   //在 SW1 上查看 OSPF 的邻居关系，它与 R1 以及 SW2 建立了
正确的邻居关系
```

```
OSPF Process 1 with Router ID 11.11.11.11
        Peer Statistic Information
 -----------------------------------------------------------------
 Area Id      Interface                    Neighbor id       State
 0.0.0.1      Vlanif8                      12.12.12.12       Full
 0.0.0.1      Vlanif9                      12.12.12.12       Full
 0.0.0.1      Vlanif10                     12.12.12.12       Full
 0.0.0.1      Vlanif11                     1.1.1.1           Full
 -----------------------------------------------------------------
<HW-SW1>display ip routing-table protocol ospf   //在 SW1 上查看 OSPF 学到的路由
Route Flags: R - relay, D - download to fib
------------------------------------------------------------------
Public routing table : OSPF
            Destinations : 2        Routes : 5

OSPF routing table status : <Active>
            Destinations : 2        Routes : 5

Destination/Mask     Proto    Pre   Cost    Flags    NextHop         Interface

    10.1.12.0/29     OSPF     10    2       D        10.1.10.12      Vlanif8
                     OSPF     10    2       D        10.1.10.29      Vlanif9
                     OSPF     10    2       D        10.1.10.44      Vlanif10
                     OSPF     10    2       D        10.1.11.1       Vlanif11
    11.1.1.1/32      OSPF     10    1       D        10.1.11.1       Vlanif11

OSPF routing table status : <Inactive>
            Destinations : 0        Routes : 0
<HW-SW2>display ospf interface

        OSPF Process 1 with Router ID 12.12.12.12
            Interfaces

 Area: 0.0.0.1
 IP Address       Type         State    Cost    Pri    DR           BDR
 10.1.10.12       Broadcast    DR       1       1      0.0.0.0      0.0.0.0
 10.1.10.29       Broadcast    DR       1       1      0.0.0.0      0.0.0.0
 10.1.10.44       Broadcast    DR       1       1      0.0.0.0      0.0.0.0
 10.1.12.2        Broadcast    DR       1       1      0.0.0.0      0.0.0.0


<HW-SW2>display ospf peer brief

        OSPF Process 1 with Router ID 12.12.12.12
            Peer Statistic Information
 -----------------------------------------------------------------
 Area Id      Interface                    Neighbor id       State
```

```
  0.0.0.1        Vlanif8                    11.11.11.11        Full
  0.0.0.1        Vlanif9                    11.11.11.11        Full
  0.0.0.1        Vlanif10                   11.11.11.11        Full
  0.0.0.1        Vlanif12                   1.1.1.1            Full
-------------------------------------------------------------------------
<HW-SW2>display ip routing-table protocol ospf    //SW2 已经通过 OSPF 学到了其他网络的路由
Route Flags: R - relay, D - download to fib
-------------------------------------------------------------------------
Public routing table : OSPF
         Destinations : 5          Routes : 14

OSPF routing table status : <Active>
         Destinations : 5          Routes : 14

Destination/Mask      Proto      Pre    Cost    Flags    NextHop        Interface

   10.1.10.14/32      OSPF       10     2       D        10.1.10.13     Vlanif8
                      OSPF       10     2       D        10.1.10.28     Vlanif9
                      OSPF       10     2       D        10.1.10.45     Vlanif10
   10.1.10.30/32      OSPF       10     2       D        10.1.10.13     Vlanif8
                      OSPF       10     2       D        10.1.10.28     Vlanif9
                      OSPF       10     2       D        10.1.10.45     Vlanif10
   10.1.10.46/32      OSPF       10     2       D        10.1.10.13     Vlanif8
                      OSPF       10     2       D        10.1.10.28     Vlanif9
                      OSPF       10     2       D        10.1.10.45     Vlanif10
   10.1.11.0/29       OSPF       10     2       D        10.1.10.13     Vlanif8
                      OSPF       10     2       D        10.1.10.28     Vlanif9
                      OSPF       10     2       D        10.1.10.45     Vlanif10
                      OSPF       10     2       D        10.1.12.1      Vlanif12
   11.1.1.1/32        OSPF       10     1       D        10.1.12.1      Vlanif12

OSPF routing table status : <Inactive>
```

30.3.2 修改 OSPF 的网络类型

OSPF 支持很多种网络类型，在以太网中默认的网络类型是广播型网络，该类型需要选举 DR 和 BDR，所以邻居建立时间中一般有 40 秒用于选举 DR/BDR，同时也会产生两类 LSA。我们可以把广播型网络修改为点到点网络，这样在一定程度上可以缩短 OSPF 的邻居建立时间，也可以减少 LSA 的数量。

在 R1 上的操作：

```
interface GigabitEthernet0/0/1
 ip address 10.1.11.1 255.255.255.248
 ospf network-type p2p    //修改设备之间互连接口的网络类型，由以太网默认的广播型网络修改为点到点网络
 ospf enable 1 area 0.0.0.1
#
```

```
interface GigabitEthernet0/0/2
 ip address 10.1.12.1 255.255.255.248
 ospf network-type p2p
 ospf enable 1 area 0.0.0.1
SW1:
interface Vlanif8
 ip address 10.1.10.13 255.255.255.240
 ospf network-type p2p
interface Vlanif9
 ospf network-type p2p
interface Vlanif10
 ospf network-type p2p
interface Vlanif11
 ospf network-type p2p
```

在 SW2 上的操作:

```
interface Vlanif8
 ospf network-type p2p
  ospf enable 1 area 0.0.0.1
#
interface Vlanif9
 ip address 10.1.10.29 255.255.255.240
 ospf network-type p2p
  ospf enable 1 area 0.0.0.1
#
interface Vlanif10
 ospf network-type p2p
  ospf enable 1 area 0.0.0.1
#
interface Vlanif12
 ospf network-type p2p
  ospf enable 1 area 0.0.0.1
```

查看修改网络类型后接口的状态:

```
<HW-SW1>display ospf interface

         OSPF Process 1 with Router ID 11.11.11.11
                 Interfaces

Area: 0.0.0.1
IP Address        Type     State     Cost     Pri     DR         BDR
10.1.10.13        P2P      P-2-P     1        1       0.0.0.0    0.0.0.0
10.1.10.28        P2P      P-2-P     1        1       0.0.0.0    0.0.0.0
10.1.10.45        P2P      P-2-P     1        1       0.0.0.0    0.0.0.0
10.1.11.2         P2P      P-2-P     1        1       0.0.0.0    0.0.0.0
```

网络类型已经被修改为点到点网络。

30.3.3 在企业网关设备上向其他交换机下发默认路由

默认路由可以是静态路由的形式,但如果已经配置了动态路由协议,那么就需要使用对应的动态路由产生默认路由,以便于访问上层或上一级网络。

```
[R1]ip route-static 0.0.0.0 0.0.0.0 GigabitEthernet0/0/0 202.100.1.2   //R1 上已经配置了静态默认路由
[R1]dis ip rou pro static    //验证前面案例中创建的静态默认路由是否存在
Route Flags: R - relay, D - download to fib
------------------------------------------------------------------------
Public routing table : Static
         Destinations : 1        Routes : 1        Configured Routes : 1

Static routing table status : <Active>
         Destinations : 1        Routes : 1

Destination/Mask    Proto      Pre    Cost    Flags    NextHop        Interface

     0.0.0.0/0     Static     60     0        D       202.100.1.2    GigabitEthernet0/0/0
[R1-ospf-1]dis th
[V200R003C00]
#
ospf 1 router-id 1.1.1.1
     default-route-advertise    //用该命令向常规区域下发 OSPF 的默认路由,默认路由以 OSPF 外部路
由的形式体现(O-ASE),该命令下发默认路由的条件是设备上已经存在一条其他路由形式的默认路由


<HW-SW1>display ip routing-table protocol ospf
Route Flags: R - relay, D - download to fib
------------------------------------------------------------------------
Public routing table : OSPF
         Destinations : 3        Routes : 6

OSPF routing table status : <Active>
         Destinations : 3        Routes : 6

Destination/Mask    Proto      Pre    Cost    Flags    NextHop        Interface
     0.0.0.0/0     O_ASE      150    1        D       10.1.11.1      Vlanif11
//其他交换机上已经存在 OSPF 形式的默认路由
    10.1.12.0/29   OSPF       10     2        D       10.1.10.12     Vlanif8
                   OSPF       10     2        D       10.1.10.29     Vlanif9
                   OSPF       10     2        D       10.1.10.44     Vlanif10
                   OSPF       10     2        D       10.1.11.1      Vlanif11
    11.1.1.1/32    OSPF       10     1        D       10.1.11.1      Vlanif11

OSPF routing table status : <Inactive>
         Destinations : 0        Routes : 0
<HW-SW1>display ip routing-table protocol ospf
Route Flags: R - relay, D - download to fib
```

```
--------------------------------------------------------------------
Public routing table : OSPF
        Destinations : 3        Routes : 6

OSPF routing table status : <Active>
        Destinations : 3        Routes : 6

Destination/Mask    Proto    Pre    Cost   Flags   NextHop       Interface
     0.0.0.0/0      O_ASE    150    1      D       10.1.11.1     Vlanif11
     10.1.12.0/29   OSPF     10     2      D       10.1.10.12    Vlanif8
                    OSPF     10     2      D       10.1.10.29    Vlanif9
                    OSPF     10     2      D       10.1.10.44    Vlanif10
                    OSPF     10     2      D       10.1.11.1     Vlanif11
     11.1.1.1/32    OSPF     10     1      D       10.1.11.1     Vlanif11

OSPF routing table status : <Inactive>
        Destinations : 0        Routes : 0
```

终端设备的数据可以到达 R1 的环回接口，该数据是根据沿途的 OSPF 明细路由进行决策后完成转发的。

终端设备的数据可以到达 R1 的互联网接口，该数据通过沿途的 OSPF 默认路由转发到达 R1。当然此时终端设备并不能和互联网的（地址）8.8.8.8 实现相互通信，通信情况如图 30-4 所示。

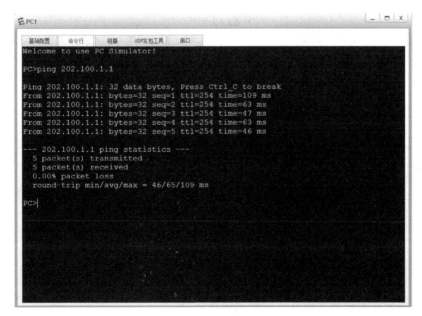

图 30-4　通信情况

本案例实施完毕。

案例 31　在思科设备上配置 OSPF

31.1　在企业内部网络上配置 OSPF

在企业内部网络上配置多区域 OSPF，以保证企业内部网络通信，同时便于企业用户网络访问互联网。

图 31-1 所示为在企业园区 AS1 内配置多区域 OSPF 的拓扑，其中 R1 的环回接口 0 位于 OSPF 区域 0（骨干区域），R1 的其他内网接口（接口 1 和接口 2），以及 SW1 和 SW2 上所有的 SVI 运行在 OSPF 区域 1。

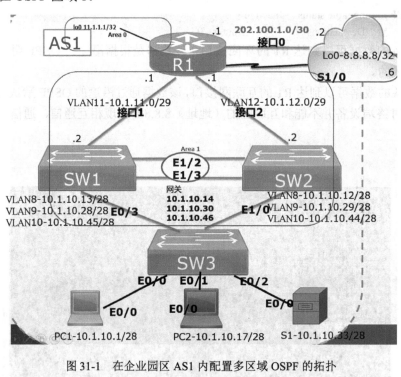

图 31-1　在企业园区 AS1 内配置多区域 OSPF 的拓扑

31.2　配置 OSPF 的要点

① 保证设备互连接口工作且直连网络通信正常。
② 创建 OSPF 进程并配置 RID。
③ 确保接口运行 OSPF 并配置正确的区域。
④ 验证和查验 OSPF 状态，完成数据通信。

31.3 配置 OSPF 的步骤详解

本实验在 R1、SW1 和 SW2 上完成，在配置 OSPF 之前需要先配置正确的 IP 地址、SVI 地址和 Trunk 链路等，以保证直连网络通信正常。

```
C-SW1#ping 10.1.11.1
Type escape sequence to abort.
Sending 5, 100-byte ICMP Echos to 10.1.11.1, timeout is 2 seconds:
!!!!!
Success rate is 100 percent (5/5), round-trip min/avg/max = 1/1/5 ms
C-SW1#ping 10.1.10.12
Type escape sequence to abort.
Sending 5, 100-byte ICMP Echos to 10.1.10.12, timeout is 2 seconds:
.!!!!
Success rate is 80 percent (4/5), round-trip min/avg/max = 1/1/1 ms
C-SW1#ping 10.1.10.29
Type escape sequence to abort.
Sending 5, 100-byte ICMP Echos to 10.1.10.29, timeout is 2 seconds:
.!!!!
Success rate is 80 percent (4/5), round-trip min/avg/max = 1/3/6 ms
C-SW1#ping 10.1.10.44
Type escape sequence to abort.
Sending 5, 100-byte ICMP Echos to 10.1.10.44, timeout is 2 seconds:
.!!!!
Success rate is 80 percent (4/5), round-trip min/avg/max = 1/1/1 ms
```

31.3.1 配置基本的 OSPF 邻居并在业务网络中运行 OSPF

R1 的配置如下：

```
R1-Cisco(config)#router ospf 110
R1-Cisco(config-router)#router-id 1.1.1.1
% OSPF: Reload or use "clear ip ospf process" command, for this to take effect   //此处演示了一个可能出现的配置场景，即如果在之前的配置中已经自动选举了一个 RID，则在重新定义时需要重启 OSPF 进程（clear ip ospf process）。注意，手动为 OSPF 指定唯一的路由器 ID
interface Loopback0
    ip address 11.1.1.1 255.255.255.255   //为了便于测试，在 R1 上创建一个用于测试的环回接口（逻辑接口），注意，环回接口在动态路由协议和高级应用中都具有非常重要的作用
    ip ospf 110 area 0   //该接口运行在 OSPF 区域 0 中，即骨干区域中
interface Ethernet0/1
    ip address 10.1.11.1 255.255.255.248
    ip ospf 110 area 1   //在接口上开启 OSPF，用于建立 OSPF 邻居，使该接口所在网络能被其他邻居学习到
    ip ospf network point-to-point   //修改默认的网络类型。在以太网上，OSPF 的网络类型默认为广播型，为了加快邻居的收敛速度，减少 LSA，推荐使用点到点的网络类型
interface Ethernet0/2
    ip address 10.1.12.1 255.255.255.248
```

```
    ip ospf 110 area 1
     ip ospf network point-to-point

C-R1#show ip ospf interface brief
Interface        PID      Area     IP Address/Mask      Cost      State Nbrs      F/C
Lo0              110      0        11.1.1.1/32          1         LOOP            0/0
Et0/2            110      1        10.1.12.1/24         10        P2P             1/1
Et0/1            110      1        10.1.11.1/29         10        P2P             1/1
```

在 SW1 和 SW2 上也运行 OSPF，手动指定各个设备的 RID。

```
C-SW1(config)#router ospf 110
C-SW1(config-router)# router-id 11.11.11.11
C-SW1(config)#interface range vlan 8 - 11
C-SW1(config-if-range)#ip ospf 110 area 1
C-SW1(config-if-range)#ip ospf network point-to-point
SW2:
router ospf 110
 router-id 12.12.12.12
!
interface Vlan8
 ip address 10.1.10.12 255.255.255.240
 ip ospf network point-to-point
 ip ospf 110 area 1
 vrrp 8 ip 10.1.10.14
!
interface Vlan9
 ip address 10.1.10.29 255.255.255.240
 ip ospf network point-to-point
 ip ospf 110 area 1
 vrrp 9 ip 10.1.10.30
 vrrp 9 priority 110
!
interface Vlan10
 ip address 10.1.10.44 255.255.255.240
 ip ospf network point-to-point
 ip ospf 110 area 1
 vrrp 10 ip 10.1.10.46
interface Vlan12
 ip address 10.1.12.2 255.255.255.248
 ip ospf network point-to-point
 ip ospf 110 area 1
```

验证运行 OSPF 的接口和 OSPF 邻居的状态。

```
C-SW1#show ip ospf interface brief
Interface        PID      Area     IP Address/Mask      Cost      State Nbrs      F/C
Vl11             110      1        10.1.11.2/29         1         P2P             1/1
Vl10             110      1        10.1.10.45/28        1         P2P             1/1
Vl9              110      1        10.1.10.28/28        1         P2P             1/1
Vl8              110      1        10.1.10.13/28        1         P2P             1/1
```

```
C-SW1#show ip ospf neighbor

Neighbor ID     Pri     State       Dead Time    Address        Interface
1.1.1.1         0       FULL/  -    00:00:39     10.1.11.1      Vlan11
12.12.12.12     0       FULL/  -    00:00:33     10.1.10.44     Vlan10
12.12.12.12     0       FULL/  -    00:00:35     10.1.10.29     Vlan9
12.12.12.12     0       FULL/  -    00:00:37     10.1.10.12     Vlan8
C-SW1#show ip route ospf
Codes: L - local, C - connected, S - static, R - RIP, M - mobile, B - BGP
       D - EIGRP, EX - EIGRP external, O - OSPF, IA - OSPF inter area
       N1 - OSPF NSSA external type 1, N2 - OSPF NSSA external type 2
       E1 - OSPF external type 1, E2 - OSPF external type 2
       i - IS-IS, su - IS-IS summary, L1 - IS-IS level-1, L2 - IS-IS level-2
       ia - IS-IS inter area, * - candidate default, U - per-user static route
       o - ODR, P - periodic downloaded static route, H - NHRP, l - LISP
       a - application route
       + - replicated route, % - next hop override

Gateway of last resort is not set

      10.0.0.0/8 is variably subnetted, 9 subnets, 3 masks
O        10.1.12.0/29 [110/11] via 10.1.11.1, 00:00:00, Vlan11
                     [110/11] via 10.1.10.44, 00:06:21, Vlan10
                     [110/11] via 10.1.10.29, 00:06:21, Vlan9
                     [110/11] via 10.1.10.12, 00:06:21, Vlan8
      11.0.0.0/32 is subnetted, 1 subnets
O IA     11.1.1.1 [110/2] via 10.1.11.1, 00:08:01, Vlan11
```

查看其他设备的 OSPF 路由。

```
C-R1#show ip route ospf
Codes: L - local, C - connected, S - static, R - RIP, M - mobile, B - BGP
       D - EIGRP, EX - EIGRP external, O - OSPF, IA - OSPF inter area
       N1 - OSPF NSSA external type 1, N2 - OSPF NSSA external type 2
       E1 - OSPF external type 1, E2 - OSPF external type 2
       i - IS-IS, su - IS-IS summary, L1 - IS-IS level-1, L2 - IS-IS level-2
       ia - IS-IS inter area, * - candidate default, U - per-user static route
       o - ODR, P - periodic downloaded static route, H - NHRP, l - LISP
       a - application route
       + - replicated route, % - next hop override

Gateway of last resort is 202.100.1.2 to network 0.0.0.0

      10.0.0.0/8 is variably subnetted, 7 subnets, 3 masks
O        10.1.10.0/28 [110/11] via 10.1.12.2, 00:01:37, Ethernet0/2    //R1 通过 OSPF 学习到了到达终
端所在网络的路由表，这就是 OSPF 最基本的作用，此时到达路由的开销相同，所以出现等价负载均衡路由
                     [110/11] via 10.1.11.2, 00:14:05, Ethernet0/1
O        10.1.10.16/28 [110/11] via 10.1.12.2, 00:01:37, Ethernet0/2
                     [110/11] via 10.1.11.2, 00:14:05, Ethernet0/1
```

```
O        10.1.10.32/28 [110/11] via 10.1.12.2, 00:01:37, Ethernet0/2
                      [110/11] via 10.1.11.2, 00:14:05, Ethernet0/1
C-SW2#show ip route ospf
Codes: L - local, C - connected, S - static, R - RIP, M - mobile, B - BGP
       D - EIGRP, EX - EIGRP external, O - OSPF, IA - OSPF inter area
       N1 - OSPF NSSA external type 1, N2 - OSPF NSSA external type 2
       E1 - OSPF external type 1, E2 - OSPF external type 2
       i - IS-IS, su - IS-IS summary, L1 - IS-IS level-1, L2 - IS-IS level-2
       ia - IS-IS inter area, * - candidate default, U - per-user static route
       o - ODR, P - periodic downloaded static route, H - NHRP, l - LISP
       a - application route
       + - replicated route, % - next hop override

Gateway of last resort is not set

      10.0.0.0/8 is variably subnetted, 9 subnets, 3 masks
O        10.1.11.0/29 [110/2] via 10.1.10.45, 00:12:40, Vlan10
                      [110/2] via 10.1.10.28, 00:12:40, Vlan9
                      [110/2] via 10.1.10.13, 00:12:40, Vlan8
      11.0.0.0/32 is subnetted, 1 subnets
O IA     11.1.1.1 [110/2] via 10.1.12.1, 00:01:42, Vlan12
```

测试终端到达地址 11.1.1.1 的数据通信情况。

```
C-PC1#ping 10.1.10.14
Type escape sequence to abort.
Sending 5, 100-byte ICMP Echos to 10.1.10.14, timeout is 2 seconds:
!!!!!
Success rate is 100 percent (5/5), round-trip min/avg/max = 1/201/1003 ms
C-PC1#ping 11.1.1.1    //通过 OSPF 方式学到的路由可使终端设备的数据到达局域网中的其他设备
Type escape sequence to abort.
Sending 5, 100-byte ICMP Echos to 11.1.1.1, timeout is 2 seconds:
!!!!!
Success rate is 100 percent (5/5), round-trip min/avg/max = 2/2/3 ms
C-PC1#traceroute 11.1.1.1 numeric
Type escape sequence to abort.
Tracing the route to 11.1.1.1
VRF info: (vrf in name/id, vrf out name/id)
  1 10.1.10.13 2 msec 2 msec 1 msec
  2 10.1.11.1 2 msec *   2 msec


C-PC2#show ip route
Default gateway is 10.1.10.30

Host               Gateway           Last Use    Total Uses    Interface
ICMP redirect cache is empty
C-PC2#ping 11.1.1.1
```

Type escape sequence to abort.
Sending 5, 100-byte ICMP Echos to 11.1.1.1, timeout is 2 seconds:
!!!!!
Success rate is 100 percent (5/5), round-trip min/avg/max = 1/202/1004 ms
C-PC2#traceroute 11.1.1.1 numeric
Type escape sequence to abort.
Tracing the route to 11.1.1.1
VRF info: (vrf in name/id, vrf out name/id)
  1 10.1.10.29 1 msec 1 msec 2 msec
  2 10.1.12.1 2 msec *   2 msec
C-Server1#show ip route
Default gateway is 10.1.10.46

Host   Gateway   Last Use   Total Uses   Interface
ICMP redirect cache is empty
C-Server1#ping 10.1.10.46
Type escape sequence to abort.
Sending 5, 100-byte ICMP Echos to 10.1.10.46, timeout is 2 seconds:
!!!!!
Success rate is 100 percent (5/5), round-trip min/avg/max = 1/201/1003 ms
C-Server1#ping 11.1.1.1
Type escape sequence to abort.
Sending 5, 100-byte ICMP Echos to 11.1.1.1, timeout is 2 seconds:
!!!!!
Success rate is 100 percent (5/5), round-trip min/avg/max = 1/2/3 ms
C-Server1#traceroute 11.1.1.1 numeric
Type escape sequence to abort.
Tracing the route to 11.1.1.1
VRF info: (vrf in name/id, vrf out name/id)
  1 10.1.10.45 1 msec 1 msec 2 msec
  2 10.1.11.1 3 msec

31.3.2 在企业网关设备上向局域网下发 OSPF 默认路由

C-R1#show ip route static
Codes: L - local, C - connected, S - static, R - RIP, M - mobile, B - BGP
      D - EIGRP, EX - EIGRP external, O - OSPF, IA - OSPF inter area
      N1 - OSPF NSSA external type 1, N2 - OSPF NSSA external type 2
      E1 - OSPF external type 1, E2 - OSPF external type 2
      i - IS-IS, su - IS-IS summary, L1 - IS-IS level-1, L2 - IS-IS level-2
      ia - IS-IS inter area, * - candidate default, U - per-user static route
      o - ODR, P - periodic downloaded static route, H - NHRP, l - LISP
      a - application route
      + - replicated route, % - next hop override

Gateway of last resort is 202.100.1.2 to network 0.0.0.0

```
S*      0.0.0.0/0 is directly connected, Ethernet0/0    //首先，在 R1 上要存在静态方式的默认路由
        8.0.0.0/32 is subnetted, 1 subnets
S       8.8.8.8 [1/0] via 202.100.1.2, Ethernet0/0
```
C-R1(config)#router ospf 110
C-R1(config-router)#default-information originate   //该命令通过静态默认路由向其他配置 OSPF 的设备下发默认路由

C-SW1#show ip route ospf
```
Codes: L - local, C - connected, S - static, R - RIP, M - mobile, B - BGP
       D - EIGRP, EX - EIGRP external, O - OSPF, IA - OSPF inter area
       N1 - OSPF NSSA external type 1, N2 - OSPF NSSA external type 2
       E1 - OSPF external type 1, E2 - OSPF external type 2
       i - IS-IS, su - IS-IS summary, L1 - IS-IS level-1, L2 - IS-IS level-2
       ia - IS-IS inter area, * - candidate default, U - per-user static route
       o - ODR, P - periodic downloaded static route, H - NHRP, l - LISP
       a - application route
       + - replicated route, % - next hop override

Gateway of last resort is 10.1.11.1 to network 0.0.0.0

O*E2   0.0.0.0/0 [110/1] via 10.1.11.1, 00:00:18, Vlan11
       10.0.0.0/8 is variably subnetted, 9 subnets, 3 masks
O         10.1.12.0/29 [110/2] via 10.1.10.44, 00:14:38, Vlan10
                       [110/2] via 10.1.10.29, 00:14:38, Vlan9
                       [110/2] via 10.1.10.12, 00:14:38, Vlan8
       11.0.0.0/32 is subnetted, 1 subnets
O IA      11.1.1.1 [110/2] via 10.1.11.1, 00:26:39, Vlan11
```
C-SW2#show ip route ospf
```
Codes: L - local, C - connected, S - static, R - RIP, M - mobile, B - BGP
       D - EIGRP, EX - EIGRP external, O - OSPF, IA - OSPF inter area
       N1 - OSPF NSSA external type 1, N2 - OSPF NSSA external type 2
       E1 - OSPF external type 1, E2 - OSPF external type 2
       i - IS-IS, su - IS-IS summary, L1 - IS-IS level-1, L2 - IS-IS level-2
       ia - IS-IS inter area, * - candidate default, U - per-user static route
       o - ODR, P - periodic downloaded static route, H - NHRP, l - LISP
       a - application route
       + - replicated route, % - next hop override

Gateway of last resort is 10.1.12.1 to network 0.0.0.0

O*E2   0.0.0.0/0 [110/1] via 10.1.12.1, 00:00:22, Vlan12
       10.0.0.0/8 is variably subnetted, 9 subnets, 3 masks
O         10.1.11.0/29 [110/2] via 10.1.10.45, 00:25:03, Vlan10
                       [110/2] via 10.1.10.28, 00:25:03, Vlan9
                       [110/2] via 10.1.10.13, 00:25:03, Vlan8
```

```
       11.0.0.0/32 is subnetted, 1 subnets
O IA        11.1.1.1 [110/2] via 10.1.12.1, 00:14:05, Vlan12
```

测试交换机得到默认路由后的通信情况，终端将到达 8.8.8.8 的数据发送到对应的主机的网关，之后通过 OSPF 路由发往下一跳 R1，R1 通过其静态明细路由将数据发往 8.8.8.8。当然，此时尚不能完成通信，但可以把数据发送到 8.8.8.8。那如何才能完成和 8.8.8.8 这个互联网设备的通信呢？请参考后续有关 NAT 的案例。

```
C-PC1#traceroute 8.8.8.8 numeric
Type escape sequence to abort.
Tracing the route to 8.8.8.8
VRF info: (vrf in name/id, vrf out name/id)
  1 10.1.10.13 2 msec 1 msec 2 msec
  2 10.1.11.1 3 msec 3 msec 3 msec
  3  *  *  *
  4  *  *  *
  5  *  *  *
  6  *  *  *
  7  *  *  *
C-PC2#traceroute 8.8.8.8
Type escape sequence to abort.
Tracing the route to 8.8.8.8
VRF info: (vrf in name/id, vrf out name/id)
  1 10.1.10.29 1 msec 2 msec 2 msec
  2 10.1.12.1 2 msec 3 msec 2 msec
  3  *  *  *
  4  *
C-Server1#traceroute 8.8.8.8 numeric
Type escape sequence to abort.
Tracing the route to 8.8.8.8
VRF info: (vrf in name/id, vrf out name/id)
  1 10.1.10.45 1 msec 1 msec 1 msec
  2 10.1.11.1 2 msec 2 msec 3 msec
  3  *  *  *
```

本案例实施完毕。

案例 32 在华为设备上配置 GRE 隧道和 EBGP

通用路由封装和 BGP 简书

GRE（Generic Routing Encapsulation，通用路由封装）协议用于对某些网络层协议（如 IP）的数据报文进行封装，使这些被封装的数据报文能够在另一个网络层协议（如 IP）中传输。封装后的数据报文在网络中传输的路径被称为 GRE 隧道。GRE 协议是 IP 的 47 号协议，GRE 隧道是虚拟的点到点的连接，其两端的设备分别对数据报进行封装及解封装。GRE 也是一种简易的 VPN 技术，可以结合 GRE over IPSec 实现安全的穿越互联网的通信，同时它可以使用动态路由协议，这是其很大的优点。

动态路由协议可按工作范围分为 IGP（Interior Gateway Protocol，内部网关协议）和 EGP（Exterior Gateway Protocol，外部网关协议）。IGP 工作在同一个 AS（Autonomous System，自治系统）内，主要用来发现和计算路由，在 AS 内提供路由信息的交换；而 EGP 工作在 AS 与 AS 之间，在 AS 间提供无环路的路由信息交换，BGP（Border Gateway Protocol，边界网关协议）则是最重要的 EGP。有人把 BGP 称为 Bloody Good Protocol，认为 BGP 是整个互联网的基石。

自治系统过去的定义是由同一个技术管理机构管理、使用统一选路策略的一些路由器的集合，现在通常指一个大型企业或者一个运营商的 BGP 的管理域。BGP 是边界网关协议，用来在 AS 之间传递路由信息，是一种增强的距离矢量路由协议，使用可靠的路由更新机制，它具有丰富的 Metric 度量方法，而且从设计上避免了环路的发生。BGP 使用 TCP 的 179 端口。

因为要建立 TCP 连接，所以两端的路由器必须知道对方的 IP 地址，可以通过直连端口、静态路由或者 IGP 来学习完成。

ISP 边界路由器知道对方的 IP 地址后，就可以尝试跟对方建立连接了。如果无法建立连接，说明对方还未激活，于是会等待一段时间再进行连接，这个过程一直重复，直到连接建立。如果 TCP 连接建立起来，两端的设备必须交换某些数据以确认对方的能力或确定自己下一步的行动，即所谓的能力交互。这个过程是必需的，因为任何支持 IP 协议栈的设备都支持 TCP 连接的建立，但不是每个支持 IP 协议栈的设备都支持 BGP，所以必须在该 TCP 连接上进行确认。

确认对方支持 BGP 后，即可进行路由表的同步。两端路由表同步完成之后，并不是立即拆除这个 TCP 连接。如果拆除此 TCP 连接，以后路由表发生改变，进行同步时就必须重新建立连接，这样需要消耗很多资源。如果保持 TCP 连接，就可以不用重新建立连接，而马上进行数据传输。

建立连接的两台设备互为对等体（PEER）。为了确保两边设备的 BGP 进程都正在运行，要求两端的设备通过该 TCP 连接周期性发送 KeepAlive 消息，以向对端确认自身还存活。

BGP 使用 TCP 作为其承载协议，提高了协议的可靠性。BGP 无须周期性发送更新消息。当路由更新时，BGP 只需要发送路由增量（增加、修改、删除的路由信息），大大减少了传播路由信息所占用的带宽，适用于在 Internet 上传播大量的路由信息。BGP 在初始化时发送所有的路由信息给 BGP 对等体，同时在本地保存已经发送给 BGP 对等体的路由信息。当本地的 BGP 收到了一条新路由信息时，与保存的已发送信息进行比较，如未发送过，则发送，如已发送过，则与已经发送的路由信息进行比较，如果新路由更优，则发送此新路由信息，同时更新已发送信息，反之则不发送。当本地 BGP 发现一条路由失效时（如对应端口失效），如果路由信息已发送过，则向 BGP 对等体发送一个撤销路由信息的消息。总之，BGP 不是每次都广播所有的路由信息，而是在初始化全部路由信息后只发送路由增量。这样保证了 BGP 在和对端通信时占用最少的带宽。

BGP 邻居包含 IBGP（AS 号码相同）和 EBGP（AS 号码不同）两种。

BGP 路由的处理过程如图 32-1 所示，BGP 路由来源包括从其他协议引入和从邻居学习两部分。为了缩小路由表规模，可以先进行路由聚合，之后路由器进行路由选择、路由发送和下发 IP 路由表等过程。BGP 可以引入其他路由协议的路由，比如直连路由、静态路由和 IGP 路由。BGP 在引入路由时支持从其他协议引入和 Network 两种方式。

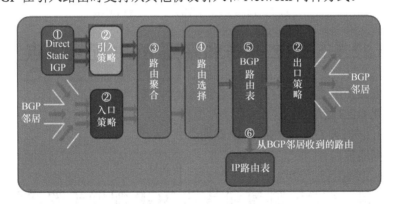

图 32-1　BGP 路由的处理过程

32.1　GRE 协议和 EBGP 应用案例说明

在企业网的两个站点间（AS1 和 AS2）的网关设备即 R1 和 R3 上配置 GRE 隧道，"打通"两个被互联网隔离的网络，之后在该隧道的基础上运行 EBGP 承载两个站点间的路由，使得两个站点间完成通信（见图 32-2）。

32.2　GRE 协议和 EBGP 应用案例拓扑

在两个站点间配置 GRE 和 EBGP，如图 32-2 所示。在两个企业网关设备 R1 和 R3 上配置 GRE 隧道，其私网地址为 10.1.13.1/30 和 10.1.13.2/30；通过在 GRE 隧道上运行 EBGP，产生 BGP 路由，使得站点间的数据穿越互联网，完成通信。

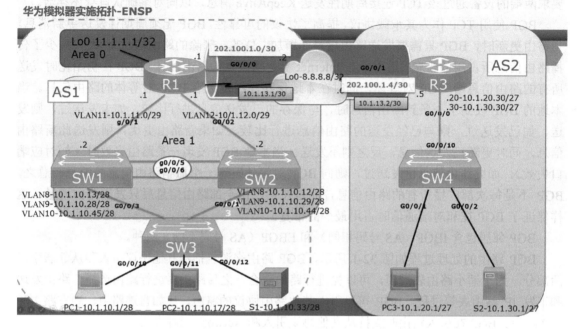

图 32-2 在两个站点间配置 GRE 和 EBGP

32.3 GRE 隧道和基础 BGP 配置要点

① 确保静态路由配置正确，GRE 隧道源地址和目的地址可达。
② 正确配置 GRE 隧道。
③ 配置 EBGP 邻居关系并产生 BGP 路由。
④ 验证数据通信。

32.4 GRE 隧道和基础 BGP 配置步骤详解

32.4.1 在华为设备上配置 GRE 隧道

确保 GRE 隧道的源地址和目的地址能在两台设备上通过稳定的路由方式学习到彼此，同时保证隧道源地址和目的地址可以通信。

```
<HW-R3>ping -a 202.100.1.5 202.100.1.1
  ping 202.100.1.1: 56   data bytes, press CTRL_C to break
    Reply from 202.100.1.1: bytes=56 Sequence=1 ttl=254 time=50 ms
    Reply from 202.100.1.1: bytes=56 Sequence=2 ttl=254 time=10 ms
    Reply from 202.100.1.1: bytes=56 Sequence=3 ttl=254 time=20 ms
    Reply from 202.100.1.1: bytes=56 Sequence=4 ttl=254 time=30 ms
    Reply from 202.100.1.1: bytes=56 Sequence=5 ttl=254 time=30 ms
```

```
--- 202.100.1.1 ping statistics ---
    5 packet(s) transmitted
    5 packet(s) received
    0.00% packet loss
    round-trip min/avg/max = 10/28/50 ms

R1
interface Tunnel0/0/0
    ip address 10.1.13.1 255.255.255.252    //GRE 隧道是私网的一部分，所以此处参照整个拓扑设计使
用了私网地址，如果不配置 IP 地址，隧道不会工作
    tunnel-protocol gre    //隧道协议使用 GRE 方式
    source 202.100.1.1    //GRE 隧道的源地址为 202.10.1.1，也可以使用接口
    destination 202.100.1.5    //GRE 隧道的目的地址是对端设备的地址，即 202.100.1.5，注意不要写
成 "discription" 而缺少了目的地址
R3
interface Tunnel0/0/0
    description 202.100.1.1    //请注意，此处为 "discription"，而非 "destination"。这是一个易犯的
错误
    ip address 10.1.13.2 255.255.255.252
    tunnel-protocol gre
    source 202.100.1.5
    destination 202.100.1.1
```

完成上述操作后，通过查看接口状态和直连路由了解隧道的状态。

```
<R1>display ip int brief
*down: administratively down
^down: standby
(l): loopback
(s): spoofing
The number of interface that is UP in Physical is 7
The number of interface that is DOWN in Physical is 1
The number of interface that is UP in Protocol is 7
The number of interface that is DOWN in Protocol is 1

Interface              IP Address/Mask    Physical    Protocol
GigabitEthernet0/0/0   202.100.1.1/30     up          up
GigabitEthernet0/0/1   10.1.11.1/29       up          up
GigabitEthernet0/0/2   10.1.12.1/29       up          up
LoopBack0              11.1.1.1/32        up          up(s)
NULL0                  unassigned         up          up(s)
Serial1/0/0            unassigned         up          up
Serial1/0/1            unassigned         down        down
Tunnel0/0/0            10.1.13.1/30       up          up    //隧道在物理和协议上处于工作状态

<R1>display ip routing-table protocol direct
Route Flags: R - relay, D - download to fib
------------------------------------------------------------------
Public routing table : Direct
         Destinations : 17        Routes : 17
```

```
Direct routing table status : <Active>
         Destinations : 17        Routes : 17

Destination/Mask    Proto      Pre   Cost  Flags  NextHop      Interface
      10.1.11.0/29  Direct     0     0     D      10.1.11.1    GigabitEthernet0/0/1
      10.1.11.1/32  Direct     0     0     D      127.0.0.1    GigabitEthernet0/0/1
      10.1.11.7/32  Direct     0     0     D      127.0.0.1    GigabitEthernet0/0/1
      10.1.12.0/29  Direct     0     0     D      10.1.12.1    GigabitEthernet0/0/2
      10.1.12.1/32  Direct     0     0     D      127.0.0.1    GigabitEthernet0/0/2
      10.1.12.7/32  Direct     0     0     D      127.0.0.1    GigabitEthernet0/0/2
      10.1.13.0/30  Direct     0     0     D      10.1.13.1    Tunnel0/0/0
      10.1.13.1/32  Direct     0     0     D      127.0.0.1    Tunnel0/0/0
      10.1.13.3/32  Direct     0     0     D      127.0.0.1    Tunnel0/0/0
       11.1.1.1/32  Direct     0     0     D      127.0.0.1    LoopBack0
      127.0.0.0/8   Direct     0     0     D      127.0.0.1    InLoopBack0
      127.0.0.1/32  Direct     0     0     D      127.0.0.1    InLoopBack0
127.255.255.255/32  Direct     0     0     D      127.0.0.1    InLoopBack0
    202.100.1.0/30  Direct     0     0     D      202.100.1.1  GigabitEthernet0/0/0
    202.100.1.1/32  Direct     0     0     D      127.0.0.1    GigabitEthernet0/0/0
    202.100.1.3/32  Direct     0     0     D      127.0.0.1    GigabitEthernet0/0/0
255.255.255.255/32  Direct     0     0     D      127.0.0.1    InLoopBack0

Direct routing table status : <Inactive>
         Destinations : 0         Routes : 0

<HW-R3>display ip int brief     //通过该命令可以看到隧道接口已经工作
*down: administratively down
^down: standby
(l): loopback
(s): spoofing
The number of interface that is UP in Physical is 6
The number of interface that is DOWN in Physical is 1
The number of interface that is UP in Protocol is 5
The number of interface that is DOWN in Protocol is 2

Interface                    IP Address/Mask        Physical    Protocol
GigabitEthernet0/0/0         unassigned             up          down
GigabitEthernet0/0/0.20      10.1.20.30/27          up          up
GigabitEthernet0/0/0.30      10.1.30.30/27          up          up
GigabitEthernet0/0/1         202.100.1.5/30         up          up
GigabitEthernet0/0/2         unassigned             down        down
NULL0                        unassigned             up          up(s)
Tunnel0/0/0                  10.1.13.2/30           up          up

<HW-R3>display ip routing-table protocol direct
Route Flags: R - relay, D - download to fib
------------------------------------------------------------
```

```
Public routing table : Direct
         Destinations : 16        Routes : 16

Direct routing table status : <Active>
         Destinations : 16        Routes : 16

Destination/Mask      Proto    Pre   Cost   Flags   NextHop        Interface
     10.1.13.0/30     Direct    0    0       D      10.1.13.2      Tunnel0/0/0
     10.1.13.2/32     Direct    0    0       D      127.0.0.1      Tunnel0/0/0
     10.1.13.3/32     Direct    0    0       D      127.0.0.1      Tunnel0/0/0
     10.1.20.0/27     Direct    0    0       D      10.1.20.30     GigabitEthernet0/0/0.20
     10.1.20.30/32    Direct    0    0       D      127.0.0.1      GigabitEthernet0/0/0.20
     10.1.20.31/32    Direct    0    0       D      127.0.0.1      GigabitEthernet0/0/0.20
     10.1.30.0/27     Direct    0    0       D      10.1.30.30     GigabitEthernet0/0/0.30
     10.1.30.30/32    Direct    0    0       D      127.0.0.1      GigabitEthernet0/0/0.30
     10.1.30.31/32    Direct    0    0       D      127.0.0.1      GigabitEthernet0/0/0.30
     127.0.0.0/8      Direct    0    0       D      127.0.0.1      InLoopBack0
     127.0.0.1/32     Direct    0    0       D      127.0.0.1      InLoopBack0
 127.255.255.255/32   Direct    0    0       D      127.0.0.1      InLoopBack0
     202.100.1.4/30   Direct    0    0       D      202.100.1.5    GigabitEthernet0/0/1
     202.100.1.5/32   Direct    0    0       D      127.0.0.1      GigabitEthernet0/0/1
     202.100.1.7/32   Direct    0    0       D      127.0.0.1      GigabitEthernet0/0/1
 255.255.255.255/32   Direct    0    0       D      127.0.0.1      InLoopBack0

Direct routing table status : <Inactive>
         Destinations : 0         Routes : 0
```

通过隧道测试逻辑上直连的对端地址的通信情况。

```
<HW-R3>ping 10.1.13.1
  ping 10.1.13.1: 56  data bytes, press CTRL_C to break
    Reply from 10.1.13.1: bytes=56 Sequence=1 ttl=255 time=30 ms
    Reply from 10.1.13.1: bytes=56 Sequence=2 ttl=255 time=30 ms
    Reply from 10.1.13.1: bytes=56 Sequence=3 ttl=255 time=40 ms
    Reply from 10.1.13.1: bytes=56 Sequence=4 ttl=255 time=30 ms
    Reply from 10.1.13.1: bytes=56 Sequence=5 ttl=255 time=30 ms

  --- 10.1.13.1 ping statistics ---
    5 packet(s) transmitted
    5 packet(s) received
    0.00% packet loss
    round-trip min/avg/max = 30/32/40 ms
```

至此，隧道配置完毕。此时，R1 和 R3 在逻辑上通过 GRE 隧道直连。接下来配置 EBGP。

32.4.2　通过配置 EBGP 邻居实现站点间通信

EBGP 通常建立在两个 AS 的直连网络上，所以在配置 EBGP 之前，我们先完成了两个 AS 之间直连的隧道接口的配置。

R1:
bgp 1   //启动 BGP，设备的 AS 号码为 1，一个设备只能启动一个 BGP 进程
[R1-bgp] peer 10.1.13.2 as-number 2   //指定对端地址 10.1.13.2，同该地址建立 EBGP 邻居关系，对端的 AS 号码为 2
[R1-bgp] import-route ospf 1   //在 R1 上，将 OSPF 路由引入 BGP 的数据库，以便其他邻居学习这些路由
!
R3:
bgp 2   //启动 BGP，设备的 AS 号码为 2
[HW-R3-bgp]peer 10.1.13.1 as-number 1   //R2 通过单播方式与 10.1.13.1 建立 EBGP 邻居，对端 AS 号码为 1
[HW-R3-bgp]   network 10.1.20.0 27   //在 R3 上通过 network 命令产生 BGP 路由，注意该命令仅用于产生路由，因而没有建立邻居的作用。采用该方式的前提是 10.1.20.0/27 已经以其他形式的路由存在于本地设备上，比如本案例中 10.1.20.0/27 以直连路由的形式存在于 R3 上，同时掩码必须精准匹配本地路由的掩码长度，否则会发生路由失效的情况。产生路由后，可以被 EBGP 邻居 R1 学习到
[HW-R3-bgp]   network 10.1.30.0 27

完成以上配置后，呈现给读者的将是下面的命令，请参考下面的命令解释：

<R1>display current-configuration conf bgp
[V200R003C00]
#
bgp 1
 peer 10.1.13.2 as-number 2
#
ipv4-family unicast   //BGP 支持多地址族，而默认的地址族是 IPv4 单播地址族
  undo synchronization
  import-route ospf 1   //将 OSPF 进程 1 引入 BGP，以使 R3 可以学习到这些路由
  peer 10.1.13.2 enable   //由于默认为 IPv4 地址族，所以自动激活了 10.1.13.2 的邻居
<HW-R3>dis cu conf bgp
[V200R003C00]
#
bgp 2
 peer 10.1.13.1 as-number 1
#
ipv4-family unicast
  undo synchronization
  network 10.1.20.0 255.255.255.224
  network 10.1.30.0 255.255.255.224
  peer 10.1.13.1 enable

验证 BGP 的邻居关系，只有存在正常的邻居关系才能更新 BGP 路由。

<HW-R3>display bgp peer

BGP local router ID : 10.1.20.30
Local AS number : 2
Total number of peers : 1   Peers in established state : 1

Peer   V   AS   MsgRcvd   MsgSent   OutQ Up/Down   State PrefRcv

10.1.13.1   4   1   27   26   0   00:21:27 Established   9

案例 32 在华为设备上配置 GRE 隧道和 EBGP

// "Established" 代表邻居关系已经成功建立,"9" 代表从对端接收到了 9 条 BGP 路由

查看 BGP 路由表,其实是 BGP 的数据库,从中可通过 BGP 选路原则得出最优路由并装载到路由表中。

```
<HW-R3>display bgp routing-table

BGP Local router ID is 10.1.20.30
Status codes: * - valid, > - best, d - damped,
              h - history,   i - internal, s - suppressed, S - Stale
              Origin : i - IGP, e - EGP, ? - incomplete


Total Number of Routes: 11
     Network            NextHop         MED         LocPrf      PrefVal Path/Ogn

 *>  10.1.10.0/28       10.1.13.1       2                       0       1?
 *>  10.1.10.14/32      10.1.13.1       2                       0       1?
 *>  10.1.10.16/28      10.1.13.1       2                       0       1?
 *>  10.1.10.30/32      10.1.13.1       2                       0       1?
 *>  10.1.10.32/28      10.1.13.1       2                       0       1?
 *>  10.1.10.46/32      10.1.13.1       2                       0       1?
 *>  10.1.11.0/29       10.1.13.1       0                       0       1?
 *>  10.1.12.0/29       10.1.13.1       0                       0       1?
 *>  10.1.20.0/27       0.0.0.0         0                       0       i
 *>  10.1.30.0/27       0.0.0.0         0                       0       i
 *>  11.1.1.1/32        10.1.13.1       0                       0       1?
```

查看通过 BGP 装载到路由表中的路由。

```
<HW-R3>display ip routing-table protocol bgp
Route Flags: R - relay, D - download to fib
------------------------------------------------------------------------------
Public routing table : BGP
         Destinations : 9        Routes : 9

BGP routing table status : <Active>
         Destinations : 9        Routes : 9

Destination/Mask    Proto    Pre   Cost      Flags NextHop         Interface

    10.1.10.0/28    EBGP     255   2           D   10.1.13.1       Tunnel0/0/0    //通过 EBGP
```

方式得到 10.1.10.0/28 的路由,EBGP 的优先级为 255,到达该路由的下一跳是 10.1.13.1,出接口为隧道 0

```
    10.1.10.14/32   EBGP     255   2           D   10.1.13.1       Tunnel0/0/0
    10.1.10.16/28   EBGP     255   2           D   10.1.13.1       Tunnel0/0/0
    10.1.10.30/32   EBGP     255   2           D   10.1.13.1       Tunnel0/0/0
    10.1.10.32/28   EBGP     255   2           D   10.1.13.1       Tunnel0/0/0
    10.1.10.46/32   EBGP     255   2           D   10.1.13.1       Tunnel0/0/0
    10.1.11.0/29    EBGP     255   0           D   10.1.13.1       Tunnel0/0/0
    10.1.12.0/29    EBGP     255   0           D   10.1.13.1       Tunnel0/0/0
    11.1.1.1/32     EBGP     255   0           D   10.1.13.1       Tunnel0/0/0
```

查看 R1 上的 BGP 邻居关系。

```
<R1>display bgp peer

BGP local router ID : 202.100.1.1
Local AS number : 1
Total number of peers : 1              Peers in established state : 1

  Peer            V        AS   MsgRcvd   MsgSent   OutQ  Up/Down        State PrefRcv

  10.1.13.2       4        2    26        28        0     00:22:15 Established   2
//邻居关系建立，而且从对端学习到 2 条 BGP 路由
```

查看通过 BGP 学习到的路由，得到了 AS2 中的两个业务网络路由：

```
<R1>display ip routing-table protocol bgp
Route Flags: R - relay, D - download to fib
------------------------------------------------------------------------------
Public routing table : BGP
         Destinations : 2        Routes : 2

BGP routing table status : <Active>
         Destinations : 2        Routes : 2

Destination/Mask      Proto    Pre  Cost       Flags   NextHop         Interface

  10.1.20.0/27        EBGP     255  0          D       10.1.13.2       Tunnel0/0/0
  10.1.30.0/27        EBGP     255  0          D       10.1.13.2       Tunnel0/0/0

BGP routing table status : <Inactive>
         Destinations : 0        Routes : 0
```

在设备双方得到各自的路由之后，进行业务网络的网关间的数据通信测试：

```
<HW-SW1>ping -a 10.1.10.14 10.1.20.30
  Warning: The specified source address is not a local address, the ping command will not check the network connection.
    ping 10.1.20.30: 56   data bytes, press CTRL_C to break
      Reply from 10.1.20.30: bytes=56 Sequence=1 ttl=254 time=50 ms
      Reply from 10.1.20.30: bytes=56 Sequence=2 ttl=254 time=50 ms
      Reply from 10.1.20.30: bytes=56 Sequence=3 ttl=254 time=40 ms
      Reply from 10.1.20.30: bytes=56 Sequence=4 ttl=254 time=60 ms
      Reply from 10.1.20.30: bytes=56 Sequence=5 ttl=254 time=70 ms

    --- 10.1.20.30 ping statistics ---
      5 packet(s) transmitted
      5 packet(s) received
      0.00% packet loss
      round-trip min/avg/max = 40/54/70 ms

<HW-SW1>ping -a 10.1.10.30 10.1.20.30
  Warning: The specified source address is not a local address, the ping command will not check the network connection.
    ping 10.1.20.30: 56   data bytes, press CTRL_C to break
      Reply from 10.1.20.30: bytes=56 Sequence=1 ttl=254 time=50 ms
      Reply from 10.1.20.30: bytes=56 Sequence=2 ttl=254 time=30 ms
```

```
    Reply from 10.1.20.30: bytes=56 Sequence=3 ttl=254 time=50 ms
    Reply from 10.1.20.30: bytes=56 Sequence=4 ttl=254 time=60 ms
    Reply from 10.1.20.30: bytes=56 Sequence=5 ttl=254 time=60 ms

  --- 10.1.20.30 ping statistics ---
    5 packet(s) transmitted
    5 packet(s) received
    0.00% packet loss
    round-trip min/avg/max = 30/50/60 ms

<HW-SW1>ping -a 10.1.10.46 10.1.20.30
  Warning: The specified source address is not a local address, the ping command will not check the
network connection.
    ping 10.1.20.30: 56    data bytes, press CTRL_C to break
    Reply from 10.1.20.30: bytes=56 Sequence=1 ttl=254 time=30 ms
    Reply from 10.1.20.30: bytes=56 Sequence=2 ttl=254 time=60 ms
    Reply from 10.1.20.30: bytes=56 Sequence=3 ttl=254 time=30 ms
    Reply from 10.1.20.30: bytes=56 Sequence=4 ttl=254 time=50 ms
    Reply from 10.1.20.30: bytes=56 Sequence=5 ttl=254 time=40 ms

  --- 10.1.20.30 ping statistics ---
    5 packet(s) transmitted
    5 packet(s) received
    0.00% packet loss
    round-trip min/avg/max = 30/42/60 ms
```

如图 32-3 所示，测试网关下终端的最终通信情况，结果显示通信正常。

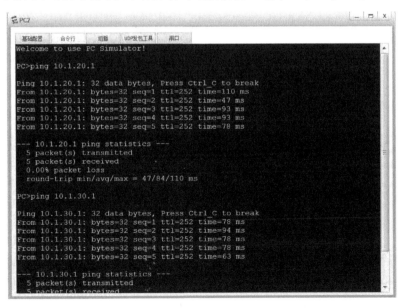

图 32-3　测试网关下终端的最终通信情况

本案例实施完毕。

案例 33　在思科设备上配置 GRE 隧道和 EBGP

33.1　GRE 协议和 EBGP 应用案例说明

在企业的两个站点（AS1 和 AS2）间配置 GRE 隧道，"打通"被互联网隔离的两个远程网络，让它们看上去在逻辑上是直连的。同时运行 EBGP，通过 BGP 路由，使得两个站点完成通信。

33.2　GRE 协议和 EBGP 应用案例拓扑

在企业网关配置 GRE 隧道和 EBGP 如图 33-1 所示。在两个企业网关设备 R1 和 R3 上配置 GRE 隧道，其私网地址为 10.1.13.1/30 和 10.1.13.2/30；通过隧道运行 EBGP，产生 BGP 路由，使得站点间实现通信。

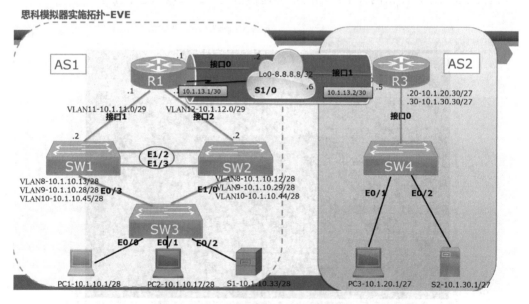

图 33-1　在企业网关配置 GRE 隧道和 EBGP

33.3　GRE 隧道和基础 BGP 配置要点

① 确保静态路由配置正确，GRE 隧道源地址和目的地址可达。

② 正确配置 GRE 隧道。
③ 配置 EBGP 邻居关系并产生 BGP 路由。
④ 验证数据通信。

33.4 GRE 隧道和基础 BGP 配置步骤详解

33.4.1 配置穿越互联网的 GRE 隧道

请自行配置 Internet（R2）设备的 IP 地址，完成企业网关设备的默认路由配置，其作用是保证隧道的源地址和目的地址互相可达。

```
C-R1(config)#ip route 0.0.0.0 0.0.0.0 Ethernet0/0 202.100.1.2
R3(config)#ip route 0.0.0.0 0.0.0.0 e0/1 202.100.1.6
```

验证静态路由，此处 R1 和 R3 之间可以通过默认路由相互访问对端的地址。

```
C-R1#show ip route static
Codes: L - local, C - connected, S - static, R - RIP, M - mobile, B - BGP
       D - EIGRP, EX - EIGRP external, O - OSPF, IA - OSPF inter area
       N1 - OSPF NSSA external type 1, N2 - OSPF NSSA external type 2
       E1 - OSPF external type 1, E2 - OSPF external type 2
       i - IS-IS, su - IS-IS summary, L1 - IS-IS level-1, L2 - IS-IS level-2
       ia - IS-IS inter area, * - candidate default, U - per-user static route
       o - ODR, P - periodic downloaded static route, H - NHRP, l - LISP
       a - application route
       + - replicated route, % - next hop override

Gateway of last resort is 202.100.1.2 to network 0.0.0.0

S*    0.0.0.0/0 [1/0] via 202.100.1.2, Ethernet0/0
R3#show ip route static
Codes: L - local, C - connected, S - static, R - RIP, M - mobile, B - BGP
       D - EIGRP, EX - EIGRP external, O - OSPF, IA - OSPF inter area
       N1 - OSPF NSSA external type 1, N2 - OSPF NSSA external type 2
       E1 - OSPF external type 1, E2 - OSPF external type 2
       i - IS-IS, su - IS-IS summary, L1 - IS-IS level-1, L2 - IS-IS level-2
       ia - IS-IS inter area, * - candidate default, U - per-user static route
       o - ODR, P - periodic downloaded static route, H - NHRP, l - LISP
       a - application route
       + - replicated route, % - next hop override

Gateway of last resort is 202.100.1.6 to network 0.0.0.0

S*    0.0.0.0/0 [1/0] via 202.100.1.6, Ethernet0/1
C-R1#ping 202.100.1.5 source 202.100.1.1
Type escape sequence to abort.
Sending 5, 100-byte ICMP Echos to 202.100.1.5, timeout is 2 seconds:
Packet sent with a source address of 202.100.1.1
```

!!!!!
Success rate is 100 percent (5/5), round-trip min/avg/max = 1/1/2 ms

在确保隧道的源地址和目的地地址之间可以通信后，再配置 GRE 隧道。

C-R1(config)#interface tunnel 0
C-R1(config-if)#
*Jan 8 08:02:42.252: %LINEPROTO-5-UPDOWN: Line protocol on Interface Tunnel0, changed state to down
C-R1(config-if)#tunnel source 202.100.1.1   //指定隧道的源 IP 地址
C-R1(config-if)#tunnel destination 202.100.1.5 //指定隧道的目的 IP 地址
C-R1(config-if)#
*Jan 8 08:02:57.913: %LINEPROTO-5-UPDOWN: Line protocol on Interface Tunnel0, changed state to up
C-R1(config-if)#ip address 10.1.13.1 255.255.255.252   //配置隧道私有网络的 IP 地址
!
R3(config)#interface tunnel 0
*Jan 8 08:03:31.003: %LINEPROTO-5-UPDOWN: Line protocol on Interface Tunnel0, changed state to down
R3(config-if)#tunnel source 202.100.1.5
R3(config-if)#tunnel destination 202.100.1.1
R3(config-if)#
*Jan 8 08:03:40.882: %LINEPROTO-5-UPDOWN: Line protocol on Interface Tunnel0, changed state to up
R3(config-if)#ip address 10.1.13.2 255.255.255.252

验证隧道接口是否工作，查看通过隧道产生的直连路由情况。

R3(config-if)#do sh ip rou static
Codes: L - local, C - connected, S - static, R - RIP, M - mobile, B - BGP
       D - EIGRP, EX - EIGRP external, O - OSPF, IA - OSPF inter area
       N1 - OSPF NSSA external type 1, N2 - OSPF NSSA external type 2
       E1 - OSPF external type 1, E2 - OSPF external type 2
       i - IS-IS, su - IS-IS summary, L1 - IS-IS level-1, L2 - IS-IS level-2
       ia - IS-IS inter area, * - candidate default, U - per-user static route
       o - ODR, P - periodic downloaded static route, H - NHRP, l - LISP
       a - application route
       + - replicated route, % - next hop override

Gateway of last resort is 202.100.1.6 to network 0.0.0.0

S*   0.0.0.0/0 [1/0] via 202.100.1.6, Ethernet0/1
      10.0.0.0/8 is variably subnetted, 2 subnets, 2 masks
R3(config-if)#do sh ip int bri
Interface   IP-Address   OK? Method Status   Protocol
Ethernet0/0   unassigned   YES NVRAM up   up
Ethernet0/1   202.100.1.5   YES NVRAM up   up
Ethernet0/2   unassigned   YES NVRAM up   up
Ethernet0/3   unassigned   YES NVRAM up   up
Tunnel0   10.1.13.2   YES manual up   up
C-R1(config-if)#do sh ip int bri
Interface   IP-Address   OK? Method Status   Protocol

Ethernet0/0	202.100.1.1	YES NVRAM	up	up
Ethernet0/1	10.1.11.1	YES NVRAM	up	up
Ethernet0/2	10.1.12.1	YES NVRAM	up	up
Ethernet0/3	unassigned	YES NVRAM	up	up
Serial1/0	unassigned	YES NVRAM	administratively down	down
Serial1/1	unassigned	YES NVRAM	administratively down	down
Serial1/2	unassigned	YES NVRAM	administratively down	down
Serial1/3	unassigned	YES NVRAM	administratively down	down
Loopback0	11.1.1.1	YES NVRAM	up	up
Tunnel0	**10.1.13.1**	**YES manual**	**up**	**up**

测试隧道直连设备的通信情况。

```
C-R1(config-if)#do ping 10.1.13.2
Type escape sequence to abort.
Sending 5, 100-byte ICMP Echos to 10.1.13.2, timeout is 2 seconds:
!!!!!
Success rate is 100 percent (5/5), round-trip min/avg/max = 1/1/2 ms
```

33.4.2 使用隧道的私网地址建立 EBGP 邻居并通信

使用隧道的私网地址（10.1.13.0/30）建立 EBGP 邻居，然后通过 BGP 得到远端网络的路由，完成站点间设备通信。

```
C-R1(config)#router bgp 1    //启动 BGP，本设备 AS 号码为 1，一台设备只能有一个 AS 进程
C-R1(config-router)# redistribute ospf 110    //在之前的实验中已经在 R1 上得到了本地局域网的路由，在这个前提下，把 OSPF 路由重分布到 BGP 中，以便对端邻居学习到这些路由
C-R1(config-router)# neighbor 10.1.13.2 remote-as 2    //通过单播方式与对端地址 10.1.13.2 建立 EBGP 邻居关系，对端设备 AS 号码为 2
C-R1#
*Jan  8 08:29:02.580: %BGP-5-ADJCHANGE: neighbor 10.1.13.2 Up    //该日志表明 BGP 邻居关系已经建立

R3(config)#router bgp 2
R3(config-router)# bgp log-neighbor-changes
R3(config-router)# neighbor 10.1.13.1 remote-as 1    //R3 通过单播方式和 R1 建立 EBGP 邻居关系，对端 AS 号码为 1
R3(config-router)#network 10.1.20.0 mask 255.255.255.224    //由于 R3 设备上业务网络较少，可以通过 network 方式产生 BGP 路由。注意，network 命令仅具备产生路由的功能，而不具备建立邻居关系的功能
R3(config-router)#network 10.1.30.0 mask 255.255.255.224
```

验证 BGP 的邻居关系，只有设备之间建立正常的 BGP 邻居关系后才可以在设备之间交互路由信息。

```
C-R1#show ip bgp summary
BGP router identifier 11.1.1.1, local AS number 1
BGP table version is 15, main routing table version 15
8 network entries using 1120 bytes of memory
8 path entries using 608 bytes of memory
3/3 BGP path/bestpath attribute entries using 420 bytes of memory
1 BGP AS-PATH entries using 24 bytes of memory
0 BGP route-map cache entries using 0 bytes of memory
```

```
0 BGP filter-list cache entries using 0 bytes of memory
BGP using 2172 total bytes of memory
BGP activity 8/0 prefixes, 9/1 paths, scan interval 60 secs

Neighbor      V    AS  MsgRcvd MsgSent   TblVer  InQ OutQ Up/Down   State/PfxRcd
10.1.13.2     4    2     20      21       15     0    0  00:14:00       2       //在状
```
态一列，如果看到数字，表示邻居关系已经建立完成，此处表示从对端收到了 2 条 BGP 路由条目
```
C-R1#
R3#show ip bgp summary
BGP router identifier 202.100.1.5, local AS number 2
BGP table version is 9, main routing table version 9
8 network entries using 1120 bytes of memory
8 path entries using 608 bytes of memory
3/3 BGP path/bestpath attribute entries using 420 bytes of memory
1 BGP AS-PATH entries using 24 bytes of memory
0 BGP route-map cache entries using 0 bytes of memory
0 BGP filter-list cache entries using 0 bytes of memory
BGP using 2172 total bytes of memory
BGP activity 8/0 prefixes, 8/0 paths, scan interval 60 secs

Neighbor      V    AS  MsgRcvd MsgSent   TblVer  InQ OutQ Up/Down   State/PfxRcd
10.1.13.1     4    1     21      20        9     0    0  00:14:10       6
```

查看学习到的具体的 BGP 路由条目：

```
R3#show ip bgp
BGP table version is 9, local router ID is 202.100.1.5
Status codes: s suppressed, d damped, h history, * valid, > best, i - internal,
              r RIB-failure, S Stale, m multipath, b backup-path, f RT-Filter,
              x best-external, a additional-path, c RIB-compressed,
Origin codes: i - IGP, e - EGP, ? - incomplete
RPKI validation codes: V valid, I invalid, N Not found

     Network           Next Hop        Metric LocPrf    Weight Path
 *>  10.1.10.0/28      10.1.13.1         11                 0 1 ?
 *>  10.1.10.16/28     10.1.13.1         11                 0 1 ?
 *>  10.1.10.32/28     10.1.13.1         11                 0 1 ?
 *>  10.1.11.0/29      10.1.13.1          0                 0 1 ?
 *>  10.1.12.0/29      10.1.13.1          0                 0 1 ?   //R3 已经从 R1 所在的 AS1 中学习到
```
了完整的 BGP 路由条目，此处的"?"表示通过重分布方式学习到的 BGP 路由
```
 *>  10.1.20.0/27      0.0.0.0            0             32768 i
 *>  10.1.30.0/27      0.0.0.0            0             32768 i
 *>  11.1.1.1/32       10.1.13.1          0                 0 1 ?
C-R1#show ip bgp
BGP table version is 15, local router ID is 11.1.1.1
Status codes: s suppressed, d damped, h history, * valid, > best, i - internal,
              r RIB-failure, S Stale, m multipath, b backup-path, f RT-Filter,
              x best-external, a additional-path, c RIB-compressed,
Origin codes: i - IGP, e - EGP, ? - incomplete
```

RPKI validation codes: V valid, I invalid, N Not found

	Network	Next Hop	Metric	LocPrf	Weight Path	
*>	10.1.10.0/28	10.1.11.2	11		32768 ?	
*>	10.1.10.16/28	10.1.11.2	11		32768 ?	
*>	10.1.10.32/28	10.1.11.2	11		32768 ?	
*>	10.1.11.0/29	0.0.0.0	0		32768 ?	
*>	10.1.12.0/29	0.0.0.0	0		32768 ?	
*>	10.1.20.0/27	10.1.13.2	0		0 2 i	
*>	10.1.30.0/27	10.1.13.2	0		0 2 i	//R1 从 AS2 的 R2 处学习到了路由，

此处的"i"表示通过 network 方式得到的路由

*>	11.1.1.1/32	0.0.0.0	0		32768 ?	

在本案例中，由于 BGP 只有一条路径，不存在冗余问题，所以 BGP 会把学习到的路由都放入路由表，如下所示：

```
C-R1#show ip route bgp
Codes: L - local, C - connected, S - static, R - RIP, M - mobile, B - BGP
       D - EIGRP, EX - EIGRP external, O - OSPF, IA - OSPF inter area
       N1 - OSPF NSSA external type 1, N2 - OSPF NSSA external type 2
       E1 - OSPF external type 1, E2 - OSPF external type 2
       i - IS-IS, su - IS-IS summary, L1 - IS-IS level-1, L2 - IS-IS level-2
       ia - IS-IS inter area, * - candidate default, U - per-user static route
       o - ODR, P - periodic downloaded static route, H - NHRP, l - LISP
       a - application route
       + - replicated route, % - next hop override

Gateway of last resort is 202.100.1.2 to network 0.0.0.0

      10.0.0.0/8 is variably subnetted, 11 subnets, 5 masks
B        10.1.20.0/27 [20/0] via 10.1.13.2, 00:10:25
B        10.1.30.0/27 [20/0] via 10.1.13.2, 00:09:54


R3#show ip route bgp
Codes: L - local, C - connected, S - static, R - RIP, M - mobile, B - BGP
       D - EIGRP, EX - EIGRP external, O - OSPF, IA - OSPF inter area
       N1 - OSPF NSSA external type 1, N2 - OSPF NSSA external type 2
       E1 - OSPF external type 1, E2 - OSPF external type 2
       i - IS-IS, su - IS-IS summary, L1 - IS-IS level-1, L2 - IS-IS level-2
       ia - IS-IS inter area, * - candidate default, U - per-user static route
       o - ODR, P - periodic downloaded static route, H - NHRP, l - LISP
       a - application route
       + - replicated route, % - next hop override

Gateway of last resort is 202.100.1.6 to network 0.0.0.0

      10.0.0.0/8 is variably subnetted, 11 subnets, 5 masks
B        10.1.10.0/28 [20/11] via 10.1.13.1, 00:14:51
```

B		10.1.10.16/28 [20/11] via 10.1.13.1, 00:14:51
B		10.1.10.32/28 [20/11] via 10.1.13.1, 00:14:51
B		10.1.11.0/29 [20/0] via 10.1.13.1, 00:14:51
B		10.1.12.0/29 [20/0] via 10.1.13.1, 00:14:51
	11.0.0.0/32 is subnetted, 1 subnets	
B		11.1.1.1 [20/0] via 10.1.13.1, 00:14:51

在两个站点建立正确的路由后，测试终端网关之间的通信情况，建议先测试终端的网关设备的通信情况，再测试最终终端的通信情况，如下所示：

```
R3#ping 10.1.10.14 source e0/0.20
Type escape sequence to abort.
Sending 5, 100-byte ICMP Echos to 10.1.10.14, timeout is 2 seconds:
Packet sent with a source address of 10.1.20.30
.!!!!
Success rate is 80 percent (4/5), round-trip min/avg/max = 2/2/3 ms
R3#ping 10.1.10.30 source e0/0.20
Type escape sequence to abort.
Sending 5, 100-byte ICMP Echos to 10.1.10.30, timeout is 2 seconds:
Packet sent with a source address of 10.1.20.30
!!!!!
Success rate is 100 percent (5/5), round-trip min/avg/max = 1/1/2 ms
R3#ping 10.1.10.46 source e0/0.20
Type escape sequence to abort.
Sending 5, 100-byte ICMP Echos to 10.1.10.46, timeout is 2 seconds:
Packet sent with a source address of 10.1.20.30
.!!!!
Success rate is 80 percent (4/5), round-trip min/avg/max = 2/2/4 ms
R3#
```

测试终端之间的通信情况：

```
Server2#ping 10.1.10.1
Type escape sequence to abort.
Sending 5, 100-byte ICMP Echos to 10.1.10.1, timeout is 2 seconds:
..!!!
Success rate is 60 percent (3/5), round-trip min/avg/max = 3/3/4 ms
Server2#ping 10.1.10.17
Type escape sequence to abort.
Sending 5, 100-byte ICMP Echos to 10.1.10.17, timeout is 2 seconds:
..!!!
Success rate is 60 percent (3/5), round-trip min/avg/max = 4/4/5 ms
Server2#ping 10.1.10.33
Type escape sequence to abort.
Sending 5, 100-byte ICMP Echos to 10.1.10.33, timeout is 2 seconds:
..!!!
Success rate is 60 percent (3/5), round-trip min/avg/max = 4/5/6 ms
```

本案例实施完毕。

第四篇　网络扩展与公网接入

基本的交换知识和路由知识并不是整个网络技术的全部内容，除此之外还需要掌握基本的互联网接入、安全管控、限速等技术，我们将在本篇中学习这些内容。很多初学者认为网络能够通信就完成了任务，但其实还需要进行许多优化与调整，本篇会为读者展示如何使用 ACL 安全地管控网络，如何使用 NAT 接入到互联网并提供服务，以及网络服务质量应用最多的限速功能。

本篇内容包含案例 34～39。

案例 34　在华为设备上配置访问控制列表

访问控制列表简书

1. ACL 的基本原理

访问控制列表（Access Control List，ACL）使用包过滤技术，在路由器上读取第三层及第四层包头中的信息，如源地址、目的地址、源端口、目的端口等信息，根据预先定义好的规则对包进行过滤，从而达到访问控制的目的。

2. ACL 的功能

网络中的节点包括资源节点和用户节点两类，其中资源节点提供服务或数据，用户节点访问资源节点所提供的服务与数据。ACL 的主要功能就是一方面保护资源节点，阻止非法用户对资源节点的访问；另一方面限制特定的用户节点所能具备的访问权限。

ACL 由一系列规则组成，通过将报文与 ACL 规则进行匹配，设备可以过滤出特定的报文、路由等，ACL 的主要动作有 permit/deny 两种，表示允许（命中）/拒绝（忽略）。

3. ACL 的分类

数字 ACL：传统的 ACL 标识方法。在创建 ACL 时，指定一个唯一的数字标识该 ACL。

命名 ACL：通过名称代替编号来标识 ACL，具有更高的识别度。

基本 ACL：华为基本 ACL 可以使用报文的源 IP 地址、分片标记和时间段信息来匹配报文，其编号取值范围为 2000~2999。思科称之为标准 ACL，其编号取值范围一般为 1~99 以及另外的扩展范围为 1300~1999，它检查源地址，匹配或者拒绝整个协议栈。

高级 ACL：可以使用报文的源/目的 IP 地址、源/目的端口号，以及协议类型等信息来匹配报文。高级 ACL 可以定义比基本 ACL 更准确、更丰富、更灵活的规则。华为设备中高级 ACL 编号的取值范围为 3000~3999。思科称之为扩展 ACL，范围为 100~199，或 2000~2699。它检查源和目标地址，匹配特定的协议或者应用。

二层 ACL：可以使用源/目的 MAC 地址以及二层协议类型等二层信息来匹配报文，华为二层 ACL 编号的取值范围为 4000~4999。思科的二层 ACL 没有规定范围。

4. ACL 报文匹配规则

一个 ACL 可以由多条 "deny | permit" 语句组成，每一条语句描述了一条规则。设备收到数据流量后，会逐条匹配 ACL 规则，看其是否匹配。如果不匹配，则匹配下一条。一旦找到一条匹配的规则，则执行规则中定义的动作，并不再继续与后续规则进行匹配。如果找不到匹配的规则，则设备不对报文进行任何处理。需要注意的是，ACL 中定义的这些规则可能存在重复或矛盾的地方。规则的匹配顺序决定了规则的优先级，ACL 通过设置规则的优先级

来处理规则之间重复或矛盾的情形。

配置顺序按 ACL 规则编号（rule-id）从小到大的顺序进行匹配。设备会在创建 ACL 的过程中自动为每一条规则分配一个编号，规则编号决定了规则被匹配的顺序。例如，如果将步长设定为 5，则规则编号将按照 5、10、15……这样的规律自动分配。如果步长设定为 2，则规则编号将按照 2、4、6、8……这样的规律自动分配。通过设置步长，使规则之间留有一定的空间，用户可以在已存在的两个规则之间插入新的规则。

通配符掩码（并不是反掩码）定义了如何检查一致的地址位，即它代表一个范围：
- 二进制的 0 代表匹配位；
- 二进制的 1 代表忽略位。

比如可以用 172.30.16.0 0.0.15.255 来表示 172.30.16.0/24 到 172.30.31.0/24 的网络。

图 34-1 所示为通配符掩码示意图。

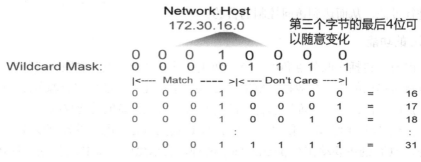

图 34-1　通配符掩码示意图

34.1　配置访问控制列表案例及拓扑

在园区网内部配置 ACL 拓扑，如图 34-2 所示。在图 34-2 中，我们将分别在网络中的物理接口和用户接口配置 ACL，用于流量过滤和用户的安全接入。在完成基本的内网通信的前提下，在网络中应用基本 ACL 和高级 ACL，可增强网络安全性和管控性。同时也方便理解数据流的方向性，以及在对应的方向上正确运用 ACL 这个常用工具。

34.2　配置 ACL 要点

① 在接口上配置基本 ACL 用于过滤流量。
② 在接口上配置高级 ACL 用于管理流量。
③ 在用户接口下调用 ACL，即 ACL 必须应用才能生效。

34.3　配置 ACL 步骤详解

34.3.1　在接口上配置基本 ACL 用于过滤流量

在 R1 上定义基本 ACL，定义完毕规则后应用于相应接口。

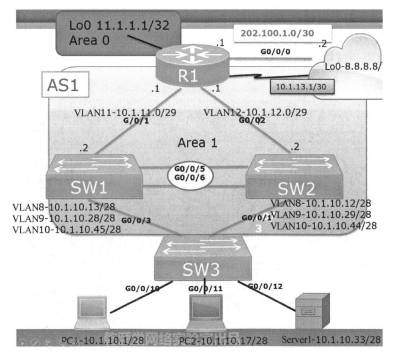

图 34-2　在园区网内部配置 ACL 拓扑

　　[R1]acl number 2000　　//进入 ACL 配置视图，ACL 命名为数字 2000
　　[R1-acl-basic-2000]rule 5 deny source 10.1.10.1 0　　//定义规则 5，拒绝来自源地址 10.1.10.1 的特定主机的所有流量
　　[R1-acl-basic-2000]int g0/0/1
　　[R1-GigabitEthernet0/0/1]traffic-filter inbound acl 2000　　//把 ACL 应用于接口，使其仅针对入方向流量生效

从 PC1 发送数据（源地址为 10.1.10.1）到 R1 的环回接口进行数据测试。

　　PC>ping 11.1.1.1　　//PC1 发送数据到 R1 的环回接口，所有 icmp 数据都被 ACL 拒绝

　　ping 11.1.1.1: 32 data bytes, Press Ctrl_C to break
　　Request timeout!
　　Request timeout!
　　Request timeout!
　　Request timeout!
　　Request timeout!

　　--- 11.1.1.1 ping statistics ---
　　　5 packet(s) transmitted
　　　0 packet(s) received
　　　100.00% packet loss
　　[R1]display acl 2000　　//验证 ACL，查看数据的匹配情况
　　Basic ACL 2000, 2 rules
　　Acl's step is 5
　　 rule 5 deny source 10.1.10.1 0 (5 matches)　　//规则 5 定义的流量有命中报文，这说明 ACL 拒绝了这类流量

使用 PC2 进行数据测试，该数据经过了 SW1，然后转发到 R1，实现了通信。请注意，华为设备的 ACL 在应用于接口时，最后一行隐含的行为是允许所有行为，这点不同于其他厂商，PC2 上的数据测试结果如图 34-3 所示。

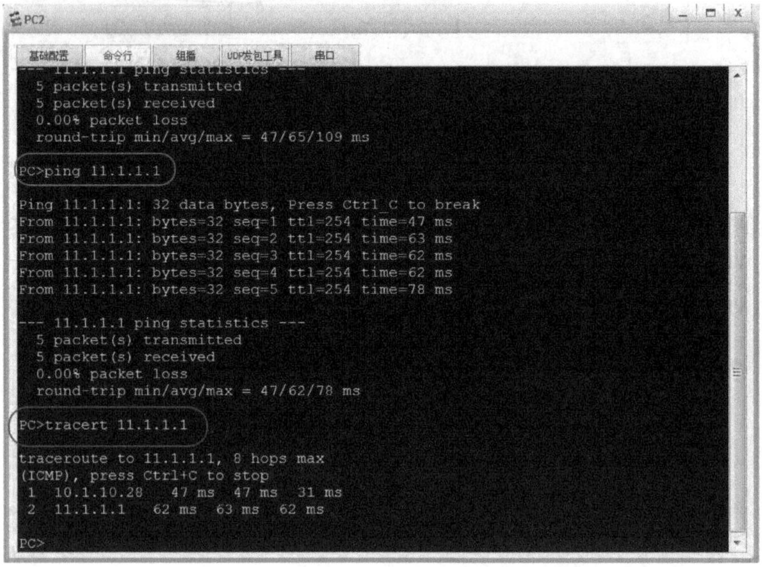

图 34-3　PC2 上的数据测试结果

建议在华为设备上增加一行命令，配置允许其他所有流量通过，方便其他工程师排查。

[R1]acl 2000
[R1-acl-basic-2000]rule 10 permit source any

使用 Server1 发送数据，服务器设备的测试如图 34-4 所示。

图 34-4　服务器设备的测试

再次验证"acl 2000"的匹配情况：

```
[R1]display acl 2000
Basic ACL 2000, 2 rules
Acl's step is 5
  rule 5 deny source 10.1.10.1 0 (5 matches)
  rule 10 permit (61 matches)   //可以看到规则10，即允许所有流量的规则命中了多次
```

34.3.2 验证 ACL 管理设备自身发起的流量

正常情况下，应用于接口的 ACL 对于从设备始发的流量不能起到管理的作用。

```
[R1]acl 2001
[R1-acl-basic-2001]rule 5 deny source 11.1.1.1 0   //拒绝从源地址 11.1.1.1 发出的流量
[R1-acl-basic-2001]rule 10 deny source 10.1.11.1 0   //拒绝从源地址 10.1.11.1 发出的流量
[R1]int g0/0/1
[R1-GigabitEthernet0/0/1]traffic-filter outbound acl 2001   //把该 ACL 应用于 G0/0/1 接口的出方向
```

测试从 R1 自身的 11.1.1.1 地址发出到达 Server1 的数据，验证数据的通信情况。

```
[R1-GigabitEthernet0/0/1]ping -a 11.1.1.1 10.1.10.33
  ping 10.1.10.33: 56  data bytes, press CTRL_C to break
    Reply from 10.1.10.33: bytes=56 Sequence=1 ttl=254 time=50 ms
    Reply from 10.1.10.33: bytes=56 Sequence=2 ttl=254 time=50 ms
    Reply from 10.1.10.33: bytes=56 Sequence=3 ttl=254 time=50 ms
    Reply from 10.1.10.33: bytes=56 Sequence=4 ttl=254 time=30 ms
    Reply from 10.1.10.33: bytes=56 Sequence=5 ttl=254 time=50 ms

  --- 10.1.10.33 ping statistics ---
    5 packet(s) transmitted
    5 packet(s) received
    0.00% packet loss
    round-trip min/avg/max = 30/46/50 ms
[R1-GigabitEthernet0/0/1]ping 10.1.10.17   //测试从 R1 的 G0/0/1 接口发出的到达地址 10.1.10.17 的
```
数据（如果不指定源，那么数据从该路由的出接口发出）
```
  ping 10.1.10.17: 56  data bytes, press CTRL_C to break
    Reply from 10.1.10.17: bytes=56 Sequence=1 ttl=127 time=60 ms
    Reply from 10.1.10.17: bytes=56 Sequence=2 ttl=127 time=60 ms
    Reply from 10.1.10.17: bytes=56 Sequence=3 ttl=127 time=50 ms
    Reply from 10.1.10.17: bytes=56 Sequence=4 ttl=127 time=50 ms
    Reply from 10.1.10.17: bytes=56 Sequence=5 ttl=127 time=70 ms

  --- 10.1.10.17 ping statistics ---
    5 packet(s) transmitted
    5 packet(s) received
    0.00% packet loss
    round-trip min/avg/max = 50/58/70 ms

[R1-GigabitEthernet0/0/1]dis acl 2001   //查看 ACL 数据匹配的情况
Basic ACL 2001, 2 rules
Acl's step is 5
```

         rule 5 deny source 11.1.1.1 0 (5 matches)   //拒绝流量的规则有流量匹配，正常情况下不能实现通信，但在模拟器上却能进行数据通信
         rule 10 deny source 10.1.11.1 0 (90 matches)   //拒绝流量的规则有流量匹配，正常情况下不能实现通信，但在模拟器上却能进行数据通信。所以，请注意 eNSP 模拟器在模拟始发于设备本身的流量时存在 bug，真机实验则没有问题

ACL 在应用到本地发起的 PBR 时是可以生效的，关于该部分内容读者可以关注笔者的其他作品。

34.3.3 将高级 ACL 应用于接口管理流量

高级 ACL 的优势在于可以精准定义源 IP 地址、目的 IP 地址、协议、端口、应用甚至字段中的一些内容，所以它的应用场景更多一些。接下来管理 R1 上的 ping 流量。

```
R1:
acl number 3000
  rule 10 deny icmp source 10.1.10.1 0 destination 11.1.1.1 0   //高级 acl 3000 拒绝源地址 10.1.10.1 到目的地址 11.1.1.1 的 ICMP（即 ping 应用）流量
  rule 20 permit ip    //显示配置规则 20 允许其他所有 IP 及 IP 层以上流量
[R1-acl-adv-3000]int g0/0/1
[R1-GigabitEthernet0/0/1]traffic-filter inbound acl 3000   //acl 3000 应用于 G0/0/1 接口的入方向
Error: A simplified ACL has been applied in this view.   //应用出错，该日志表明，已经有一个 ACL 被应用于接口
[R1-GigabitEthernet0/0/1]undo  traffic-filter inbound   //去掉之前入方向应用的 acl 2000，请注意，原则上一个接口的一个方向只能定义一个 ACL
[R1-GigabitEthernet0/0/1]traffic-filter inbound acl 3000   //正常应用
```

测试来自 PC1 到达地址 11.1.1.1 的流量，ICMP 流量已经被拒绝，但 PC1 发送到同一设备的地址 202.100.1.1 的数据却可以通信。PC1 上的数据测试结果如图 34-5 所示。

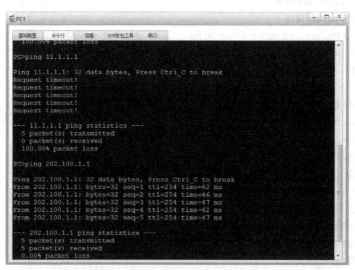

图 34-5   PC1 上的数据测试结果

在 VRP 上查看 ACL 命中流量的情况，它可以表明 ACL 生效与否。

[R1]dis acl 3000

```
Advanced ACL 3000, 2 rules
Acl's step is 5
 rule 10 deny icmp source 10.1.10.1 0 destination 11.1.1.1 0 (5 matches)
 rule 20 permit ip (238 matches)
```

接口下的应用实例演示完毕，接下来对用户接口进行 ACL 管控。

34.3.4　将 ACL 应用于用户接口实现远程管理

请参考之前的配置，在 SW1 上开启远程管理功能，此时 R1 已经可以远程管理 SW1。

```
<R1>telnet 10.1.11.2
  Press CTRL_] to quit telnet mode
  Trying 10.1.11.2 ...
  Connected to 10.1.11.2 ...

Login authentication


Password:
Info: The max number of VTY users is 5, and the number
      of current VTY users on line is 1.
      The current login time is 2019-01-09 15:40:28.
<HW-SW1>
```

配置 ACL，将其应用到 VTY 接口，使得只有那些被允许的设备才可以进行远程管理。

```
[HW-SW1]acl 3001
  [HW-SW1-acl-adv-3001]rule 5 permit tcp source 10.1.10.12 0 destination-port   eq 23    //高级 acl 3001
```
允许源地址为 10.1.10.12 的设备的 TCP 目的端口为 23 的流量，即仅允许 10.1.10.12 来远程管理 SW1
```
  [HW-SW1]user-interface vty 0 4
  [HW-SW1-ui-vty0-4]acl 3001 inbound    //把 acl 3001 应用于 VTY 接口
```

进行数据测试以验证 ACL 的管控行为。

```
<R1>telnet 10.1.11.2   //R1 已经无法管理 SW1
  Press CTRL_] to quit telnet mode
  Trying 10.1.11.2 ...

<HW-SW2>telnet 10.1.11.2    //SW2 可以使用源地址 10.1.10.12 连接目的端口 23 进行远程管理，这
```
表明 ACL 管控已经生效
```
  Trying 10.1.11.2 ...
  Press CTRL+K to abort
  Connected to 10.1.11.2 ...

Login authentication


Password:
Info: The max number of VTY users is 5, and the number
      of current VTY users on line is 1.
      The current login time is 2019-01-09 15:47:48.
```

```
<HW-SW1>
[HW-SW1]display users
    User-Intf    Delay      Type    Network Address    AuthenStatus    AuthorcmdFlag
  + 0    CON 0    00:00:00                                                              no
Username : Unspecified

    34     VTY 0    00:07:23    TEL    10.1.10.12        pass             no
Username : Unspecified    //来自地址 10.1.10.12 的用户已经成功登录了 SW1
    [HW-SW1]display acl 3001    //查看 acl 3001，读者可能会有疑惑，测试数据明明成功，ACL 却没
有任何匹配数。这是在华为 eNSP 模拟器上比较常见的 bug
    Advanced ACL 3001, 2 rules
    Acl's step is 5
      rule 5 permit tcp source 10.1.10.12 0 destination-port eq telnet
      rule 10 deny ip
```

ACL 的应用场景非常多，在本书后面的案例中还会介绍将 ACL 用于 NAT 和 QoS 的情况。本案例实施完毕。

案例 35　在思科设备上配置访问控制列表

35.1　配置访问控制列表案例及拓扑

在完成基本的内网通信后，在网络中应用标准 ACL 和扩展 ACL，可以增强网络管理的安全性和管控性。请注意，ACL 的应用非常灵活，很多内容需要管理员根据网络需求进行调整，对具体问题具体分析。在园区网内配置 ACL 管控网络拓扑如图 35-1 所示。在网络中的物理接口和用户接口配置 ACL，用于流量过滤和用户的安全接入，请确保已经通过采用 OSPF 使得整个 AS1 内部实现通信。

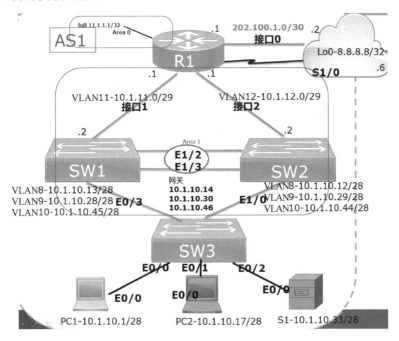

图 35-1　在园区网内配置 ACL 管控网络拓扑

35.2　配置 ACL 要点

① 在接口上配置标准 ACL 用于过滤流量，请关注通配符掩码。
② 在接口上配置扩展 ACL 用于管理流量，请关注通配符掩码。
③ 在 VTY 接口上配置 ACL，请关注通配符掩码。

35.3 配置 ACL 步骤详解

35.3.1 在接口上配置标准 ACL 用于过滤流量

在案例 31 中,我们已经在企业网内部配置 OSPF 以保证企业网内部通信,请在此前提下实施本案例。

```
C-R1(config)#ip access-list standard ?   //此处的帮助命令表明了 ACL 的数字范围
  <1-99>        Standard IP access-list number
  <1300-1999>   Standard IP access-list number (expanded range)
  WORD          Access-list name    //该选项表明在思科设备上 ACL 也可以使用命名方式
C-R1(config)#ip access-list standard 10
C-R1(config-std-nacl)#deny host 10.1.10.1   //拒绝来自主机 10.1.10.1 的所有流量,此处没有定义序号,系统会指定序号,也可以自行配置序号
C-R1(config-std-nacl)#permit any   //允许其他所有流量,否则其他流量将会被拒绝
C-R1(config)#int e0/1
C-R1(config-if)#ip access-group 10 in   //把 ACL 10 应用到接口的入方向,ACL 不应用便不会生效
```

应用 ACL 后,通过数据流量查验 ACL 生效与否。

```
C-PC1#ping 202.100.1.1
Type escape sequence to abort.
Sending 5, 100-byte ICMP Echos to 202.100.1.1, timeout is 2 seconds:
U.U.U
Success rate is 0 percent (0/5)
C-PC1#ping 11.1.1.1   //PC1 发送数据到地址 11.1.1.1,该数据无法正常通信,因为该流量被 ACL 拒绝
Type escape sequence to abort.
Sending 5, 100-byte ICMP Echos to 11.1.1.1, timeout is 2 seconds:
U.U.U
Success rate is 0 percent (0/5)
C-PC1#
C-PC2#ping 202.100.1.1
Type escape sequence to abort.
Sending 5, 100-byte ICMP Echos to 202.100.1.1, timeout is 2 seconds:
!!!!!
Success rate is 100 percent (5/5), round-trip min/avg/max = 2/2/4 ms
C-PC2#
C-PC2#ping 11.1.1.1   //PC2 发送数据到地址 11.1.1.1,该数据可以正常通信
Type escape sequence to abort.
Sending 5, 100-byte ICMP Echos to 11.1.1.1, timeout is 2 seconds:
!!!!!
Success rate is 100 percent (5/5), round-trip min/avg/max = 2/2/2 ms


C-R1#show access-lists 10   //验证 ACL 动作是否命中
Standard IP access list 10
    10 deny    10.1.10.1 (16 matches)   //代表行为"拒绝"的序号 10 的 ACL 已经命中,这表明该 ACL 表项生效
    20 permit any (35 matches)
```

标准 ACL 应用案例实施完毕，可以查看管控结果。

35.3.2　验证 ACL 管理设备自身发起的流量

在正常情况下，ACL 一般无法管控设备自身发起的流量，请参考如下案例。

```
C-R1(config)#access-list 11 deny host 11.1.1.1    //尝试拒绝从主机 11.1.1.1 发出的流量
C-R1(config)#access-list 11 deny 10.1.11.0 0.0.0.7    //尝试拒绝从 10.1.11.0/29 网络发出的流量
C-R1(config)#access-list 11 permit any    //允许其他流量
C-R1(config)#int e0/1
C-R1(config-if)#ip access-group 11 out    //该 ACL 应用到接口 1 的出方向
```

应用 ACL 后，通过 R1 发送数据来测试从设备自身发出的流量能否被 ACL 所管理。

```
C-R1#ping 10.1.10.17 source 11.1.1.1
Type escape sequence to abort.
Sending 5, 100-byte ICMP Echos to 10.1.10.17, timeout is 2 seconds:
Packet sent with a source address of 11.1.1.1
!!!!!
Success rate is 100 percent (5/5), round-trip min/avg/max = 2/2/3 ms
C-R1#ping 10.1.10.17
Type escape sequence to abort.
Sending 5, 100-byte ICMP Echos to 10.1.10.17, timeout is 2 seconds:
!!!!!
Success rate is 100 percent (5/5), round-trip min/avg/max = 2/2/3 ms
C-R1#show access-list 11    //查看 ACL，发现没有任何匹配，这表明当 ACL 应用到接口时不能管
控从设备自身发出的流量
Standard IP access list 11
    10 deny    11.1.1.1
    20 deny    10.1.11.0, wildcard bits 0.0.0.7
    30 permit any
```

正如以上案例所示，大部分时候 ACL 无法管控设备自身发出的流量。当 ACL 被应用到本地发起的 PBR 时是可以生效的，该部分内容读者可以关注笔者的其他作品。

35.3.3　将扩展 ACL 应用于接口管理流量

本节我们来配置思科的扩展 ACL，从而更加精细地控制流量。

```
C-R1(config)#ip access-list extended Traffic    //定义扩展 ACL，名字为 Traffic，命名形式更易于识
别，此处也可以使用数字形式命名
C-R1(config-ext-nacl)#deny icmp 10.1.10.1 0.0.0.0 11.1.1.1 0.0.0.0    //拒绝从 10.1.10.1 到 11.1.1.1 的
所有 ICMP（即 ping 应用）流量
C-R1(config-if)#int e0/1
C-R1(config-if)# ip access-group Traffic in    //在接口 1 应用 ACL，发现后配置的 ACL 直接应用成功
C-R1(config-ext-nacl)#permit ip any any    //允许其他所有流量
C-R1(config-if)#do sh run int e0/1    //查看接口的配置，可以看到后应用的 ACL 直接覆盖了之前配
置的 ACL，而且该接口的出、入方向各自配置了一个 ACL
Building configuration...

Current configuration : 171 bytes
```

```
!
interface Ethernet0/1
  ip address 10.1.11.1 255.255.255.248
  ip access-group Traffic in
  ip access-group 11 out
  ip ospf network point-to-point
  ip ospf 110 area 1
```

通过终端发送数据测试 ACL 管控是否成功。

```
C-PC1#ping 11.1.1.1    //从 PC1 的 10.1.10.1 到 11.1.1.1 的 ICMP 流量已经被拒绝转发
Type escape sequence to abort.
Sending 5, 100-byte ICMP Echos to 11.1.1.1, timeout is 2 seconds:
U.U.U
Success rate is 0 percent (0/5)
C-PC1#ping 202.100.1.1    //从 PC1 的 10.1.10.1 到 R1 的其他接口,比如 202.100.1.1 可以匹配到"允
许其他流量"的规则,所以可以通信
Type escape sequence to abort.
Sending 5, 100-byte ICMP Echos to 202.100.1.1, timeout is 2 seconds:
!!!!!
Success rate is 100 percent (5/5), round-trip min/avg/max = 2/2/3 ms
C-Server1#ping 11.1.1.1    //其他设备到 11.1.1.1 的流量由于源地址没有被命中,所以也被允许
Type escape sequence to abort.
Sending 5, 100-byte ICMP Echos to 11.1.1.1, timeout is 2 seconds:
!!!!!
Success rate is 100 percent (5/5), round-trip min/avg/max = 2/203/1009 ms
C-Server1#ping 202.100.1.1
Type escape sequence to abort.
Sending 5, 100-byte ICMP Echos to 202.100.1.1, timeout is 2 seconds:
!!!!!
Success rate is 100 percent (5/5), round-trip min/avg/max = 1/1/2 ms
```

请读者自行用 show access-list 命令查看命中情况。

35.3.4　将 ACL 应用于 VTY 接口实现远程管理

除了物理接口的流量,VTY 接口(即逻辑接口)也可以被 ACL 管控。

请参考之前的案例,开启 SW1 的远程管理功能,如下所示为 R1 可以成功地远程管理 SW1。

```
C-R1#telnet 10.1.10.13
Trying 10.1.10.13 ... Open


User Access Verification

Password:
C-SW1>
//为了增强网络安全,仅允许 10.1.10.12 的主机远程管理 SW1
C-SW1(config)#ip access-list extended VTY    //配置扩展访问控制列表,命名为 VTY
```

C-SW1(config-ext-nacl)#permit tcp host 10.1.10.12 any eq 23   //仅允许源地址为 10.1.10.12 的主机向目标端口 23 发送（即 telnet）的流量。这意味着其他流量被拒绝
C-SW1(config)#line vty 0 4
C-SW1(config-line)#access-class VTY in   //把 ACL 应用到 VTY 接口的入方向

再次从其他设备尝试远程管理 SW1，此时只有 SW2 可以成功实现远程管理。

```
C-R1#telnet 10.1.10.13    //R1 发起的 telnet 流量已经被拒绝
Trying 10.1.10.13 ...
% Connection refused by remote host

C-R1#

C-SW2#telnet 10.1.10.13    //SW2 可以继续发起 telnet 管理
Trying 10.1.10.13 ... Open

User Access Verification

Password:
C-SW1>

C-SW1#show users    //查看登录 SW1 的用户，可以看到只有地址为 10.1.10.12 的设备成功登录
    Line       User       Host(s)           Idle         Location
*  0 con 0               idle              00:00:00
   2 vty 0               idle              00:00:05     10.1.10.12

   Interface   User       Mode              Idle         Peer Address
```

本案例实施完毕。

案例 36 在华为设备上配置 NAT 功能

NAT 技术简书

随着互联网的发展和网络应用的增多，IPv4 地址枯竭已成为制约网络发展的瓶颈。尽管 IPv6 可以从根本上解决 IPv4 地址空间不足的问题，但目前众多网络设备和网络应用大多是基于 IPv4 的，因此在 IPv6 广泛应用之前，一些过渡技术（如 CIDR、私网地址等）的使用是解决这个问题最主要的技术手段。NAT（Network Address Translation，网络地址转换）是将 IP 数据包报文头中的 IP 地址转换为另一个 IP 地址的过程，主要用于实现内部网络（简称内网，使用私有 IP 地址）访问外部网络（简称外网，使用公有 IP 地址）的功能。当内网的主机要访问外网时，通过 NAT 技术可以将其私网地址转换为公网地址，可以实现多个私网用户共用一个公网地址来访问外部网络，这样既可以保证网络互通，又节省了公网地址。

华为 NAT 大体上可分为以下几类。
- 静态 NAT 又称 Basic NAT（静态一对一转换），思科称之为静态转换；
- 动态 NAT 又称地址池 NAT（多对多转换），思科称之为动态转换；
- Easy IP，（多对一转换），思科称之为端口地址转换（PAT）。
- 静态转换中可以实现某个特定协议（比如 TCP）的转换，即静态地址端口转换，华为称之为 NAT 服务器。

36.1 配置 NAT 功能案例及拓扑

在企业网关 R1 上配置 NAT 功能，如图 36-1 所示。在企业网关设备（R1）上实现 Easy IP 技术和 NAT 服务器功能，使得部分网络可以接入互联网，在 R2(云)上使用 8.8.8.8 模拟了一个互联网地址，在 SW1 上开启 Telnet 服务并对外提供该服务，使得 SW1 可以向互联网设备（云）提供 Telnet 服务。

36.2 配置 NAT 功能要点

① 定义正确的 ACL，ACL 命中的流量可以访问互联网，默认其他流量会被拒绝。
② 在企业网关连接互联网的接口上配置 Easy IP 功能。
③ 在企业网关上配置 NAT 服务器功能对外提供服务。
④ 使用正确的方式完成测试。

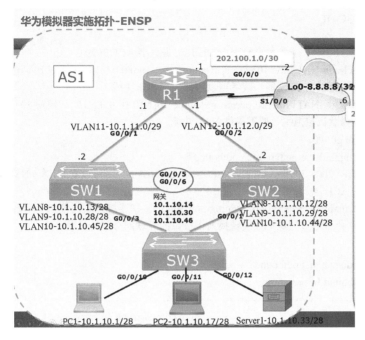

图 36-1 在企业网关 R1 上配置 NAT 功能

36.3 配置 NAT 功能步骤详解

36.3.1 配置 Easy IP 功能实现内网访问互联网

前置条件：完成内网对应 IGP、交换部分的配置，同时在企业网关（R1）上存在静态默认路由，如下所示。

```
<HW-R1>display ip routing-table protocol static
Route Flags: R - relay, D - download to fib
------------------------------------------------------------------------
Public routing table : Static
         Destinations : 1       Routes : 1       Configured Routes : 1

Static routing table status : <Active>
         Destinations : 1       Routes : 1

Destination/Mask    Proto    Pre  Cost    Flags    NextHop        Interface

      0.0.0.0/0    Static    60   0        D      202.100.1.2    GigabitEthernet0/0/0

Static routing table status : <Inactive>
         Destinations : 0       Routes : 0
```

配置 Easy IP 功能：

　　[HW-R1]acl name NAT   //配置 ACL，此处使用命名 ACL。在华为设备上命名 ACL 其实也是数字 ACL

　　[HW-R1-acl-adv-NAT]dis th

```
[V200R003C00]
#
acl name NAT 3999    //名为 NAT 的 ACL 其实是数字 ACL 3999
[HW-R1-acl-adv-NAT] rule 10 permit ip source 10.1.10.0 0.0.0.15    //规则 10 匹配源自 10.1.10.0/28 网
络的数据，对这些流量提供 NAT 功能，读者可以根据实际需求完成
[HW-R1-acl-adv-NAT] rule 20 permit ip source 10.1.10.16 0.0.0.15    //规则 20 匹配源自 10.1.20.0/28
网络的数据，其他数据默认被拒绝，即不提供 NAT 功能
[HW-R1]int g0/0/0
[HW-R1-GigabitEthernet0/0/0]nat outbound ?
    INTEGER<2000-3999>    Apply basic or advanced ACL    //可以看到，在 eNSP 模拟器上实现 NAT
转换时只能使用数字 ACL
[HW-R1-GigabitEthernet0/0/0]nat outbound 3999    //在连接互联网接口上实现 Easy IP 功能，调用
ACL 3999，ACL3999 命中流量的源地址被转换为该接口的互联网公网地址
```

代码验证：

```
<HW-R1>display nat outbound
 NAT Outbound Information:
 --------------------------------------------------------------------------
  Interface                    Acl        Address-group/IP/Interface    Type
 --------------------------------------------------------------------------
  GigabitEthernet0/0/0         3999       202.100.1.1                   easyip
 --------------------------------------------------------------------------
  Total : 1
```

流量验证：被允许的内网网络，即 PC1 和 PC2 所在的网络发起流量的源地址被转换为公网地址去访问互联网。配置 NAT 功能后业务网络访问互联网情况如图 36-2 所示。

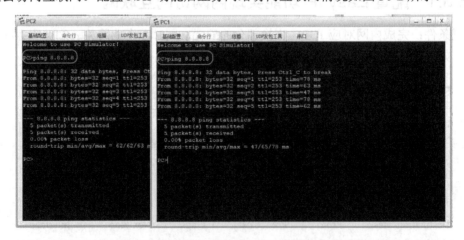

图 36-2 配置 NAT 功能后业务网络访问互联网情况

该流量的地址转换可以通过以下命令来验证：

```
<HW-R1>display nat session all
 NAT Session Table Information:

   Protocol          : ICMP(1)        //流量为 ICMP 流量
   SrcAddr    Vpn    : 10.1.10.17     //源自地址 10.1.10.17 的数据
```

```
        DestAddr    Vpn         : 8.8.8.8    //到达目的地址 8.8.8.8 的流量
        Type Code IcmpId        : 0     8     61989
        NAT-Info
          New SrcAddr           : 202.100.1.1   //10.1.10.17 的源地址被转换为 202.100.1.1
          New DestAddr          : ----
          New IcmpId            : 10250

        Protocol                : ICMP(1)
        SrcAddr     Vpn         : 10.1.10.17
        DestAddr    Vpn         : 8.8.8.8
        Type Code IcmpId        : 0     8     61991
        NAT-Info
          New SrcAddr           : 202.100.1.1
          New DestAddr          : ----
          New IcmpId            : 10252
```

而源自 Server1 的流量没有被 ACL 命中，所以不会进行地址转换，数据的源地址和目的地址依旧为 10.1.10.33→8.8.8.8，由于 R1 上存在默认路由，所以数据可以发送到互联网设备 8.8.8.8，但是互联网设备不存在（也不能存在）到企业内部网络的路由（即 10.1.10.32/28），所以数据通信不成功。图 36-3 为不能成功接入互联网示意图。

图 36-3 不能成功接入互联网示意图

36.3.2 配置 NAT 服务器功能向互联网提供服务

接下来在 R1 上配置 NAT 服务器功能，通过该功能把内网的 Telnet 服务器映射到公网，使得公网设备可以登录内部设备。

```
[HW-R1]int g0/0/0
    [HW-R1-GigabitEthernet0/0/0] nat server protocol tcp global current-interface 2323 inside 10.1.10.14
    telnet    //在连接互联网的接口上开启 NAT 服务器功能，把内部网络地址 10.1.10.14 的 23 端口（即
Telnet 服务）转换为当前接口地址的 2323 端口，这意味着当外网用户访问地址 202.100.1.1 的 2323 端口时，
会转换为地址 10.1.10.14 的 23 端口。请注意不要在 "global" 命令后使用该接口的地址，否则会报错，如下
```

所示

```
[HW-R1-GigabitEthernet0/0/0]nat server protocol tcp global 202.100.1.1 1000 inside 10.1.10.13 23
Error: The address conflicts with interface or ARP IP.
```

验证 NAT 服务器的配置结果：

```
[HW-R1]display nat server
  Nat Server Information:
    Interface    : GigabitEthernet0/0/0    //配置 NAT 接口
      Global IP/Port    : current-interface/2323 (Real IP : 202.100.1.1)    //公网地址和端口，使用当前接口（真实 IP 地址为 202.100.1.1），端口为 2323
      Inside IP/Port    : 10.1.10.14/23(telnet)    //转换到内部网络的地址为 10.1.10.14，提供的服务为 23 号服务
      Protocol : 6(tcp)    //协议为 IP 协议 6，即 TCP 协议
      VPN instance-name : ----
      Acl number        : ----
      Description : ----

  Total :    1
```

请参考前面的讲解，自行开启 SW1 的 Telnet 服务，之后在互联网设备上进行测试：

```
<Internet>telnet 202.100.1.1 2323    //互联网设备登录网关公网地址的 2323 端口
Press CTRL_] to quit telnet mode
Trying 202.100.1.1 ...
Connected to 202.100.1.1 ...

Login authentication


Password:
Info: The max number of VTY users is 5, and the number
      of current VTY users on line is 1.
      The current login time is 2019-01-10 15:38:08.
<HW-SW1>    //已经成功登录 SW1，注意此时登录的是企业内部设备，而非网关设备，即已经实现了内网的 23 端口服务


<HW-SW1>display users    //在 SW1 上查看哪些用户登录了设备，可以看到 202.100.1.2 的外网用户已经通过 Telnet 成功登录
    User-Intf   Delay    Type    Network Address    AuthenStatus    AuthorcmdFlag
  + 0    CON 0    00:00:00                                                        no
   Username : Unspecified

    34   VTY 0    00:00:08    TEL    202.100.1.2          pass           no
   Username : Unspecified


<HW-R1>display nat session protocol tcp    //在登录 SW1 的情况下，在 R1 上查看 TCP 协议的 NAT 会话
  NAT Session Table Information:

   Protocol         : TCP(6)
   SrcAddr   Port Vpn : 202.100.1.2       7104    //流量的源端口和源地址
```

```
        DestAddr Port Vpn : 202.100.1.1      4873    //流量的目的地址和目的端口
        NAT-Info
          New SrcAddr        : ----
          New SrcPort        : ----
          New DestAddr       : 10.1.10.14    //目的地址从 202.100.1.1 转换为 10.1.10.14
          New DestPort       : 5888    //被转换后的目的端口，这些端口并不是真正的端口，真正的端
口可以在交换机上查看

        Total : 1

        <HW-SW1>display tcp status    //查看 SW1 上 TCP 状态
        TCPCB     Tid/Soid Local Add:port      Foreign Add:port     VPNID   State
        16b8bcc0 77 /1    0.0.0.0:23           0.0.0.0:0            -1      Listening
        1546ce30 77 /4    10.1.10.14:23        202.100.1.2:49392    0       Established   //从互联网设
备上发起的会话，源地址为 202.100.1.2，源端口为 49392
```

本案例实施完成。

案例 37 在思科设备上配置 NAT 功能

37.1 配置 NAT 功能案例

在企业网关上配置 PAT 和静态 NAT 功能如图 37-1 所示，在企业网关 R1 上，通过采用 PAT 功能使得可控的内网用户（VLAN8 和 VLAN9）接入互联网，通过采用静态 NAT 功能使得内网服务器对外提供服务。

37.2 配置 NAT 功能案例拓扑

如图 37-1 所示，在企业网关设备上实现 Easy IP 和 NAT 服务器功能，使得部分网络可以接入互联网，在 R2（云）上使用 8.8.8.8 模拟了一个互联网地址；在 SW1 上开启 Telnet 服务并对外提供该服务，使得外网设备（云）可以访问 SW1 的 Telnet 服务。

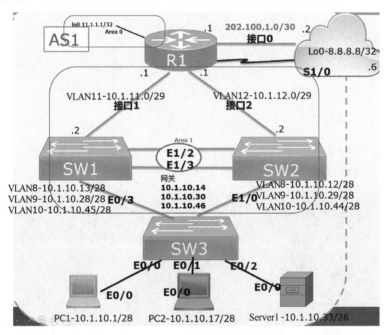

图 37-1 在企业网关上配置 PAT 和静态 NAT 功能

37.3 配置 NAT 功能要点

① 定义正确的 ACL，ACL 命中的流量可以访问互联网，默认其他流量会被拒绝。
② 在企业网关连接互联网的端口上配置 PAT 功能，注意 inside 和 outside 端口。

③ 在企业网关上配置地址端口转换功能对外提供服务。
④ 使用正确的方式完成测试。

37.4 配置 NAT 功能步骤详解

37.4.1 配置 PAT 功能实现对应内网访问互联网业务

前置条件：在网关设备 R1 上已经存在静态默认路由以便把数据发往互联网。

```
C-R1#show ip route static
Codes: L - local, C - connected, S - static, R - RIP, M - mobile, B - BGP
       D - EIGRP, EX - EIGRP external, O - OSPF, IA - OSPF inter area
       N1 - OSPF NSSA external type 1, N2 - OSPF NSSA external type 2
       E1 - OSPF external type 1, E2 - OSPF external type 2
       i - IS-IS, su - IS-IS summary, L1 - IS-IS level-1, L2 - IS-IS level-2
       ia - IS-IS inter area, * - candidate default, U - per-user static route
       o - ODR, P - periodic downloaded static route, H - NHRP, l - LISP
       a - application route
       + - replicated route, % - next hop override

Gateway of last resort is 202.100.1.2 to network 0.0.0.0

S*      0.0.0.0/0 [1/0] via 202.100.1.2, Ethernet0/0
```

在企业网关设备 R1 上配置 PAT（端口地址转换）功能。

```
    C-R1(config)#ip access-list standard NAT    //将 ACL 命名为 NAT，对它命中的流量可以提供 NAT
功能，而不被命中的流量不提供 NAT 功能，即不能访问互联网
    C-R1(config-std-nacl)#permit 10.1.10.0 0.0.0.15    //对命中 10.1.10.0/28 网络的主机可以提供 NAT 功能
    C-R1(config-std-nacl)#permit 10.1.10.16 0.0.0.15    //对命中 10.1.10.16/28 网络的主机可以提供 NAT
功能
    C-R1(config-std-nacl)#exit
    C-R1(config)#interface range e0/1 - 2
    C-R1(config-if-range)#ip nat inside    //把内网端口 e0/1-2 定义为提供 NAT 功能的内部端口，这表明，
流量必须从这些端口进入网关设备才可能实现 NAT 功能

    *Jan  9 09:16:39.996: %LINEPROTO-5-UPDOWN: Line protocol on Interface NVI0, changed state to
    up    //在定义完毕 NAT 端口后，IOS 上会启动一个 NAT 虚拟端口 0，该端口会借用连接互联网的
地址，即把内网数据转发到该地址。可以通过"show interfaces"命令查看到这个端口
    C-R1(config-if-range)#int e0/0
    C-R1(config-if)#ip nat outside    //把连接互联网的端口定义为 NAT 的 outside 端口
    C-R1(config-if)#exit
    C-R1(config)#ip nat inside source list NAT interface e0/0 overload    //配置 NAT 功能，即把符合 ACL
规则的数据的源地址转换成 e0/0 端口的公网地址（202.100.1.1）。"overload" 代表复用，即一个地址可以重
复使用多个端口（端口代表应用）
    C-R1(config)#
```

通过从多个客户端发起到互联网地址 8.8.8.8（默认用互联网设备的环回接口）的访问验证 NAT 功能是否配置成功。

C-PC1#ping 8.8.8.8   //PC1 发起流量的源地址被 ACL 命中，其源地址 10.1.10.1 被转换为 202.100.1.1，然后以源地址 202.100.1.1 到目的地址 8.8.8.8 的流量被转发到互联网；当互联网设备回送报文时，源地址为 202.100.1.2，目的地址为 202.100.1.1；当数据到达 R1 后，查看 NAT 表项，只会把目的地为 202.100.1.1 的流量转发到对应的内部地址

```
Type escape sequence to abort.
Sending 5, 100-byte ICMP Echos to 8.8.8.8, timeout is 2 seconds:
!!!!!
Success rate is 100 percent (5/5), round-trip min/avg/max = 2/2/4 ms
C-PC1#
C-PC2#ping 8.8.8.8    //PC2 发起流量的源地址被 ACL 命中，可以实现 NAT 功能，所以可以通信
Type escape sequence to abort.
Sending 5, 100-byte ICMP Echos to 8.8.8.8, timeout is 2 seconds:
!!!!!
Success rate is 100 percent (5/5), round-trip min/avg/max = 2/2/4 ms
C-PC2#
```

C-Server1#ping 8.8.8.8   //Server1 发起流量的源地址不能被 ACL 命中，所以该地址不能被转换为公网地址 202.100.1.1。此时，R1 上存在默认路由可以把源地址 10.1.10.33 到 8.8.8.8 的流量直接转发到互联网设备，但是互联网没有局域网的私网路由，即没有到 10.1.10.32/28 的路由，所以不能实现通信

```
Type escape sequence to abort.
Sending 5, 100-byte ICMP Echos to 8.8.8.8, timeout is 2 seconds:
.....
Success rate is 0 percent (0/5)
C-Server1#

C-R1#show access-list NAT    //查看 ACL 匹配情况，读者可以看到相关流量被 ACL 命中
Standard IP access list NAT
    10 permit 10.1.10.0, wildcard bits 0.0.0.15 (1 match)
    20 permit 10.1.10.16, wildcard bits 0.0.0.15 (1 match)
C-R1#show ip nat translations    //查看 NAT 情况
Pro Inside global      Inside local       Outside local       Outside global
icmp 202.100.1.1:4     10.1.10.1:4        8.8.8.8:4           8.8.8.8:4
```
//10.1.10.1 的内部主机地址可以被转换为全局（即公网）地址 202.100.1.1。注意此处的":4"代表一个模拟端口号（ICMP 是 IP 协议，不存在端口号，只有真正的应用才会存在端口号）

```
icmp 202.100.1.1:2     10.1.10.17:2       8.8.8.8:2           8.8.8.8:2
```

此时部分可管控的主机访问互联网成功。

37.4.2　配置静态 NAT 功能以对外提供服务

在思科设备上定义外部 NAT 的 inside 和 outside 端口后，可以直接实现静态地址端口转换功能。

C-R1(config)#ip nat inside source static tcp 10.1.10.13 23 202.100.1.1 2323   //使用静态 NAT 功能，把内网设备 10.1.10.13（SW1）的 23 端口（即 Telnet 服务），转换为互联网地址 202.100.1.1 的 2323 端口，使得外网用户可以访问互联网地址 202.100.1.1 的 2323 端口，得到内网 10.1.10.13 提供的 23 端口服务

查看 NAT 表项：

```
C-R1#show ip nat translations
```

```
Pro Inside global        Inside local      Outside local      Outside global
tcp 202.100.1.1:2323     10.1.10.13:23     ---                ---
C-R1#
```

请从外网访问内部设备提供的服务，注意端口号为 2323，而非默认的 23 端口。

```
R2-internet#telnet 202.100.1.1 2323    //此时 Telnet 服务不正确，访问被拒绝
Trying 202.100.1.1, 2323 ...
% Connection refused by remote host
```
//其实地址转换是成功的，但是由于之前案例中配置了 VTY 被 ACL 管控的功能，所以需要修改 ACL 配置

```
C-R1#debug ip nat    //读者可以通过开启调试命令查看地址转换情况，可以看到源地址 202.100.1.2
```
到目的地址 202.100.1.1 的流量；在 R1 上，目的地址为 10.1.10.14，所以此时的流量为 202.100.1.2→10.1.10.14 的流量

```
*Jan  9 10:12:06.546: NAT*: TCP s=17811, d=2323->23
*Jan  9 10:12:06.546: NAT*: s=202.100.1.2, d=202.100.1.1->10.1.10.14 [60473]
*Jan  9 10:12:06.546: NAT*: TCP s=17811, d=2323->23
*Jan  9 10:12:06.546: NAT*: s=202.100.1.2, d=202.100.1.1->10.1.10.14 [60474]
```
//在 SW1 上修改 ACL 配置，允许通过源 202.100.1.1 的 VTY 流量
```
C-SW1(config)#ip access-list extended VTY
C-SW1(config-ext-nacl)# permit tcp host 202.100.1.2 any eq telnet
```

再次测试 Telnet 服务：

```
R2-internet#telnet 202.100.1.1 2323    //互联网设备已经可以远程管理交换机 SW1

C-SW1(config)#ip access-list extended VTY
C-SW1(config-ext-nacl)# permit tcp host 202.100.1.2 any eq telnet
Trying 202.100.1.1, 2323 ... Open

User Access Verification

Password:
C-SW1>
```

本案例实施完毕。

案例 38　在华为设备上实施网络 QoS 限速

QoS 和限速技术简书

很多网络对带宽、延迟、延迟抖动、丢包率等传输性能有一定要求，这时可以在网络中实施 QoS（Quality of Service，服务质量）技术，对不同流量采取不同策略，即"可管理的不公平"。

QoS 有以下三种服务模型：
- 尽力而为（Best-Effort）服务模型；
- 综合服务（Integrated Service）模型，简称 IntServ 模型；
- 区分服务（Differentiated Service）模型，简称 DiffServ 模型。

最实用的是区分服务模型，而实施区分服务模型通常会采用 MQC（模块化 QoS）配置方式。

QoS 业务可以分为以下几类。

① 流分类和标记（Traffic Classification and Marking）：要实现区分服务，需要首先将数据包分为不同的类别或者设置为不同的优先级。将数据包分为不同的类别称为流分类，流分类并不修改原来的数据包。将数据包设置为不同的优先级称为标记，而标记会修改原来的数据包。

② 流量监管和整形（Traffic Policing and Shaping）：指将业务流量限制在特定的带宽内，当业务流量超过额定带宽时，超过的流量将被丢弃或缓存。其中，将超过的流量丢弃的技术称为流量监管，将超过的流量缓存的技术称为流量整形。

③ 拥塞管理和避免（Congestion Management and Avoidance）：拥塞管理是指在网络发生拥塞时，将报文放入队列中缓存，并采取某种调度算法安排报文的转发次序。而拥塞避免可以监督网络资源的使用情况，当发现拥塞有加剧的趋势时采取主动丢弃报文的策略，通过调整流量来解除网络的过载问题。

思科技术中还包含链路效率，华为技术中包含端口镜像，是指对进入设备的流量进行监控，确保其没有滥用网络资源。通过监控进入网络的某一流量的规格，将它限制在一个允许的范围之内，若某个连接的报文流量过大，就对流量进行惩罚，比如丢弃报文，或重新设置该报文的优先级（比如限制 HTTP 报文不能占用超过 50%的网络带宽），以保护网络资源和运营商的利益不受损害。

流量监管采用承诺访问速率（Committed Access Rate，CAR）来对流量进行控制。CAR 使用令牌桶算法对流量速率进行评估，依据评估结果，实施预先设定好的监管动作。对应于 SLA（Service-Level Agreement，服务等级协议）预定的处理动作，流量监管动作如下：

- 转发（Pass）：对测量结果不超过承诺信息速率（Committed Information Rate，CIR）的报文，通常的处理方法为继续正常转发。
- 丢弃（Discard）：对测量结果超过峰值信息速率（Peak Information Rate，PIR）的报文通常丢弃。
- 重标记（Remark）：对处于 CIR 与 PIR 之间的流量通常执行 Remark 动作，此时的报文不丢弃，而是通过 Remark 降低优先级进行尽力而为转发。

CAR 主要有以下两个功能：
- 流量速率限制：通过使用令牌桶对流经端口的报文进行度量，使得在特定时间内只有得到令牌的流量通过，从而实现限速功能。
- 流分类：通过令牌桶算法对流量进行测量，根据测量结果给报文打上不同的流分类，内部标记（包括服务等级与丢弃优先级）。

流量监管主要应用于网络边缘入口处，对超出 SLA 约定的流量报文给予通过、直接丢弃或降级处理，从而保证在 SLA 约定范围之内的报文可享受 SLA 预定的服务，同时保证核心设备的正常数据处理。监管实施位置如图 38-1 所示，在路由器的入方向实施了监管，使得高速链路不对低速链路形成冲击。

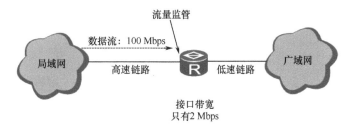

图 38-1　监管实施位置

流量整形用于对输出报文的速率进行控制，使报文以均匀的速率发送出去。进行流量整形通常是为了使报文速率与下游设备相匹配。当从高速链路向低速链路传输数据或发生突发流量时，带宽会在低速链路出口处出现瓶颈，导致数据丢失严重。在这种情况下，需要在进入低速链路的设备出口处进行流量整形。

流量整形通常使用缓冲区和令牌桶来完成，当报文的发送速度过快时，首先在缓冲区进行缓存，在令牌桶的控制下再均匀地发送这些被缓冲的报文，所以一般延时较长。

流量整形是在队列调度之后，数据包在出队列的过程中进行的。流量整形如图 38-2 所示。

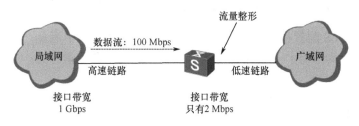

图 38-2　流量整形

38.1 实现企业网络限速管控案例及拓扑

限速是现网中常见的 QoS 实施要求。在企业网关设备 R1 上实施网络限速，如图 38-3 所示。在企业网关设备 R1 上，通过 MQC（模块化 QoS）实施监管和整形以实现网络限速，实现精准的、有保障的、可管理的网络品质。

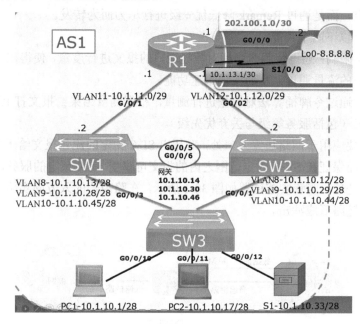

图 38-3 在企业网关设备 R1 上实施网络限速

38.2 配置 QoS 要点

① 对流量进行正确的分类是 QoS 中的难点，本例中使用 ACL 实现。
② 配置流量分类策略和流量行为策略。
③ 配置流策略并正确应用。
④ 测试和观察 QoS 限速。

38.3 配置 QoS 步骤详解

38.3.1 在华为设备上实施监管功能以实现限速

依据 MQC，QoS 配置按照以下步骤实施：① 完成 ACL 命中流量；② 配置流量分类策略；③ 配置流量行为策略；④ 使用流量策略整合流量分类和流量行为；⑤ 在接口下应用流量策略。

```
[HW-R1]acl name XIANSU    //使用 ACL 对流量分类，注意，配置 QoS 的难点就在于如何对流量
进行分类，可以根据自己的需要来定义流量，ACL 是常用方式之一
    [HW-R1-acl-adv-XIANSU] rule 10 permit ip source 10.1.10.16 0.0.0.15    //采用 ACL 定义来自
```

10.1.10.16/28 网络的流量，即对这些流量进行限速，而对其他流量不限速

```
[HW-R1-acl-adv-XIANSU]quit
[HW-R1]traffic classifier C1            //进入流分类视图，自定义的名称为 C1
[HW-R1-classifier-C1] if-match acl XIANSU    //如果命中 ACL 的流量，则进行后续动作
[HW-R1-classifier-C1]quit
[HW-R1]traffic behavior Police          //流行为视图，自定义的名称为 Police
[HW-R1-behavior-Police] car cir 8       //这是 CIR（承诺信息速率，即限定速度）参数，此处为 8 kbps，
```
其他参数 VRP 会自动配置。注意不同设备的单位不同，读者可以通过帮助命令查看。8 kbps 是 VRP 系统最小速率，是为了方便演示实验而设定的，工程师可以根据实际情况调整

```
[HW-R1-behavior-Police] statistic enable   //开启状态统计功能以便于查看实施情况
[HW-R1-behavior-Police]quit
[HW-R1]traffic policy Police    //进入流量策略视图，自定义名称为 Police，它的作用在于融合流分
```
类和流行为策略

```
[HW-R1-trafficpolicy-Police] classifier C1 behavior Police    //对流分类 C1 采用 Police 定义的流行为，
```
即限速 8 kbps

```
[HW-R1-trafficpolicy-Police]quit
[HW-R1]int g0/0/1
R1-GigabitEthernet0/0/1]traffic-policy Police inbound    //把流量监管应用到网关设备接口的入方向，
```
监管可以用到出方向和出方向

实施完监管后，通过发送流量进行验证：

```
<HW-SW1>ping -s 1000 -a 10.1.10.30 8.8.8.8    //请在 SW1 上设定源地址（请联想一下 ACL），设置
```
发送报文大小为 1000 字节，发送到 8.8.8.8，设置报文大小是为了使得设备产生超标的流量（这种设置方案并不精准，但可以达到实验效果）

```
Warning: The specified source address is not a local address, the ping command will not check the network connection.
  ping 8.8.8.8: 1000   data bytes, press CTRL_C to break
    Reply from 8.8.8.8: bytes=1000 Sequence=1 ttl=254 time=40 ms
    Request time out    //部分报文因为超过设定的速率而被丢弃
    Reply from 8.8.8.8: bytes=1000 Sequence=3 ttl=254 time=50 ms
    Request time out
    Reply from 8.8.8.8: bytes=1000 Sequence=5 ttl=254 time=30 ms

  --- 8.8.8.8 ping statistics ---
    5 packet(s) transmitted
    3 packet(s) received
    40.00% packet loss
    round-trip min/avg/max = 30/40/50 ms

<HW-SW1>
<HW-R1>display traffic policy statistics interface g0/0/1 inbound    //查看接口的流量策略状态结果，
```
可以看到部分超速流量被丢弃

```
  Interface: GigabitEthernet0/0/1
  Traffic policy inbound: Police
  Rule number: 3
  Current status: OK!
  Item                    Sum(Packets/Bytes)        Rate(pps/bps)
```

```
 -------------------------------------------------
     Matched                10/            0/
                            5,820          0
       +--Passed            8/             0/
                            3,688          0
       +--Dropped           2/             0/
                            2,132          0
         +--Filter          0/             0/
                            0              0
         +--CAR             2/             0/
                            2,132          0        //有两个报文，即 2132 字节的报文被丢弃
       +--Queue Matched     0/             0/
                            0              0
       +--Enqueued          0/             0/
                            0              0
       +--Discarded         0/             0/
                            0              0
     +--Car                 10/            0/
                            5,820          0
       +--Green packets     8/             0/
                            3,688          0
       +--Yellow packets    0/             0/
                            0              0
       +--Red packets       2/             0/
                            2,132          0
 <HW-R1>
```

采用监管方式的限速实施完毕。

38.3.2　在华为设备上实施流量整形功能以实现限速

我们依旧沿用之前的流量分类方式，即用前文定义的 C1 来实施整形策略。

```
[HW-R1-GigabitEthernet0/0/1]undo traffic-policy inbound   //先去掉之前配置的监管策略
traffic behavior P1    //定义新的流量行为，名称为 P1
  statistic enable    //开启统计功能
  gts cir 8    //自定义通用流量整形，速率设定为 8 kbps，其他参数 VRP 会自动配置
traffic policy P1
  classifier C1 behavior P1    //在流量策略中，对 C1 的分类，调用 P1 策略，即以整形方式限速。注意，在一个流量策略下可以有多个流行为
interface GigabitEthernet0/0/0
  ip address 202.100.1.1 255.255.255.252
  traffic-policy P1 outbound    //整形方式只能应用在接口的出方向
```

实施完整形策略后，请沿用之前的测试方式在 SW1 上继续验证，以此来查验整形和监管的区别。

```
<HW-SW1>ping -s 1000 -a 10.1.10.30 8.8.8.8      //依旧使用之前的测试方式。数据并没有被丢弃
Warning: The specified source address is not a local address, the ping command will not check the network connection.
  ping 8.8.8.8: 1000    data bytes, press CTRL_C to break
```

```
Reply from 8.8.8.8: bytes=1000 Sequence=1 ttl=254 time=50 ms
Reply from 8.8.8.8: bytes=1000 Sequence=2 ttl=254 time=40 ms
Reply from 8.8.8.8: bytes=1000 Sequence=3 ttl=254 time=30 ms
Reply from 8.8.8.8: bytes=1000 Sequence=4 ttl=254 time=40 ms
Reply from 8.8.8.8: bytes=1000 Sequence=5 ttl=254 time=280 ms

--- 8.8.8.8 ping statistics ---
  5 packet(s) transmitted
  5 packet(s) received
  0.00% packet loss
  round-trip min/avg/max = 30/88/280 ms

<HW-SW1>
[HW-R1]display traffic policy statistics interface g0/0/0 outbound    //查看 G0/0/0 接口出方向的流量策
略状态

Interface: GigabitEthernet0/0/0
Traffic policy outbound: P1
Rule number: 3
Current status: OK!
 Item                  Sum(Packets/Bytes)    Rate(pps/bps)
---------------------------------------------------------------
 Matched               115/                  1/
                       20,391                1,016
  +--Passed            115/                  1/
                       20,391                1,016    //所有报文都被发送，没有数据被丢弃
  +--Dropped           0/                    0/
                       0                     0
    +--Filter          0/                    0/
                       0                     0
    +--CAR             0/                    0/
                       0                     0
  +--Queue Matched     13/                   1/
   (Shaping Active:NO) 9,018                 976    //整形机制存在一个队列，该队列可以把超
速的流量放到队列中缓存，而不会丢弃数据，所以数据没有被丢弃。当然，如果队列已满，不能缓存流量，
数据报文也会被丢弃
    +--Enqueued        13/                   1/
                       9,018                 976
    +--Discarded       0/                    0/
                       0                     0
  +--Car               0/                    0/
                       0                     0
    +--Green packets   0/                    0/
                       0                     0
    +--Yellow packets  0/                    0/
                       0                     0
    +--Red packets     0/                    0/
                       0                     0
```

使用下面的验证方式验证缓存的报文超出队列之后报文被丢弃的情况：

<HW-SW1>ping -s 1501 -a 10.1.10.30 8.8.8.8   //设定较大的报文，此处为1501字节
Warning: The specified source address is not a local address, the ping command will not check the network connection.
  ping 8.8.8.8: 1501   data bytes, press CTRL_C to break
    Request time out
    Request time out
    Request time out
    Request time out
    Request time out

  --- 8.8.8.8 ping statistics ---
    5 packet(s) transmitted
    0 packet(s) received
    100.00% packet loss
<HW-R1>dis traffic policy statistics interface g0/0/0 outbound

 Interface: GigabitEthernet0/0/0
 Traffic policy outbound: P1
 Rule number: 3
 Current status: OK!

 Item   Sum(Packets/Bytes)   Rate(pps/bps)

 Matched   147/   0/
                         30,969   0
   +--Passed   147/   0/
                         30,969   0
   +--Dropped   0/   0/
                         0   0
     +--Filter   0/   0/
                         0   0
     +--CAR   0/   0/
                         0   0
   +--Queue Matched   23/   0/
   (Shaping Active:NO)   17,143   0
     +--Enqueued   18/   0/
                         16,708   0
     +--Discarded   5/   0/   //发送的5个报文全部丢弃
                         435   0
     +--Car   0/   0/
                         0   0
       +--Green packets 0/   0/
                         0   0
       +--Yellow packets 0/   0/
                         0   0
       +--Red packets   0/   0/
                         0   0

流量整形机制实施完毕。

案例 39 在思科设备上实施网络 QoS 限速

39.1 实现企业网络限速管控案例及拓扑

在企业网关设备 R1 上实现限速，如图 39-1 所示。在企业网络可以进行互联网访问的基础上，在企业网关设备（R1）上实施监管和整形以实现网络限速，对特定流量进行限制，同时保证重要数据的转发。

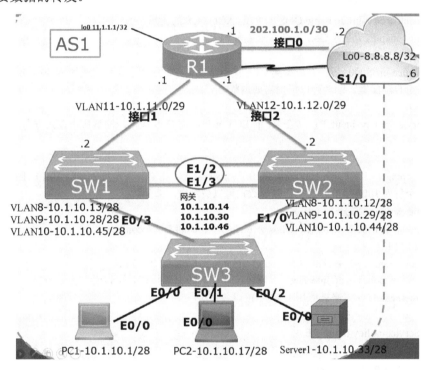

图 39-1 在企业网关设备 R1 上实现限速

39.2 配置 QoS 要点

① 对流量进行正确的分类是 QoS 中的难点，本例中使用 ACL 实现。
② 配置流量分类策略。
③ 配置流策略并正确应用。
④ 测试和观察 QoS 限速。

39.3 配置 QoS 步骤详解

39.3.1 在思科设备上实施监管功能以实现限速

依据 MQC（模块化 QoS），QoS 配置按照以下步骤实施：① 定义 ACL 命中流量；② 配置流量分类；③ 使用流量策略整合流量分类和流量行为；④ 在接口下应用流量策略。

```
R1:
  ip access-list extended QOS    //配置扩展访问扩展列表，ACL 是常用的流量分类工具，也可以使用标准 ACL
    permit ip host 10.1.10.1 any  //匹配来自主机 10.1.10.1 到任意目的的流量，对这些流量进行限速
  class-map match-all C1   //配置流量分类工具，匹配参数为 all，即需要满足该分类下所有条件，也可以调整为匹配任意，即满足该分类下的任一条件
    match access-group name QOS    //匹配 ACL QOS 的流量
  policy-map P1 //配置 QoS 策略，命名为 P1
    class C1
      police cir 8000    //针对流量类 C1，使用监管工具，配置承诺信息速率为 8000 bps，即 8 kbps，其他参数 IOS 会自动配置，读者可以根据需要调整
  int e0/1
    service-policy input P1    //在 R1 的 E0/1 接口应用 QoS 策略，注意应用在入方向，也可应用在接口的出方向

PC1#ping 11.1.1.1 size 1000 repeat 10    //依据 ACL，在 PC1（地址为 10.1.10.1）上向 R1 的一个环回地址发送每个包为 1000 字节的数据，部分数据可以被发送，部分数据将被丢弃
Type escape sequence to abort.
Sending 5, 1000-byte ICMP Echos to 11.1.1.1, timeout is 2 seconds:
!.!.!!!!!!
Success rate is 60 percent (3/5), round-trip min/avg/max = 2/2/3 ms
R1#show policy-map interface    //查看应用 QoS 策略的接口，可以看到流量情况
 Ethernet0/1    //策略被应用于 E0/1 接口

  Service-policy input: P1

    Class-map: C1 (match-all)
      10 packets, 5640 bytes
      5 minute offered rate 0000 bps, drop rate 0000 bps
      Match: access-group name QOS
      police:
          cir 8000 bps, bc 1500 bytes    //在监管策略中 CIR 被定义为 8 kbps
        conformed 8 packets, 3612 bytes; actions:    //在 CIR 内的报文有 8 个，这些报文被传输出去
          transmit
        exceeded 2 packets, 2028 bytes; actions:    //超出 CIR 的报文有 2 个，这些报文被丢弃
          drop
        conformed 0000 bps, exceeded 0000 bps
```

思科设备的监管功能实施完毕，可以根据实际情况自行调整策略和分类。

39.3.2 在思科设备上实施流量整形功能以实现限速

在原有配置的基础上继续实施 QoS 的整形策略。

  R1(config)#ip access-list extended QOS
  R1(config-ext-nacl)#permit ip host 10.1.10.1 any //使用 ACL 匹配分类的流量，读者可以根据实际情况调整，该 ACL 不会命中数据，如果沿用之前的配置，在实现 NAT 功能的情况下，源地址会被转化为 202.100.1.1，故而不会成功
  R1(config-ext-nacl)#permit ip 10.1.10.32 0.0.0.15 any //在 ACL 中增加一个新的表项，在表项中定义了 Server1 所在网络，但该网络没有配置 NAT 功能，所以其地址不会被转换成公网地址，因此 Server1 不能访问互联网，也不能 ping 通互联网设备。当然源于该网络的数据可以依据默认路由从 R1 的互联网接口发送出去
  R1(config-ext-nacl)#class-map match-all C1 //类似前面的配置，实施流量分类，默认匹配该分类的所有条件
  R1(config-cmap)#match access-group name QOS //匹配条件为匹配 ACL 流量，如果没有匹配策略，则使用默认的数据转发策略
  R1(config-cmap)#policy-map P1 //配置策略工具，命名为 P1
  R1(config-pmap)# class C1
  R1(config-pmap-c)#shape average 8000 //针对流量分类 C1，配置整形策略，设置速率为 8 kbps
  R1(config-pmap-c)#interface Ethernet0/0
  R1(config-if)# ip address 202.100.1.1 255.255.255.252
  R1(config-if)# service-policy output P1 //在接口 E0/0 的出接口调用策略

为了演示数据通信的过程，可以临时在 Internet 设备中增加一条静态路由：

  R2-internet(config)#ip route 10.1.10.32 255.255.255.240 202.100.1.1

使用 ping 命令来测试 QoS，请使用较大的报文尺寸，该方式并不精准，但可以看到一定的实验效果。

  C-Server1#ping 202.100.1.2 size 1500 repeat 10
  Type escape sequence to abort.
  Sending 10, 1500-byte ICMP Echos to 202.100.1.2, timeout is 2 seconds:
  !!!!!!!!!!
  Success rate is 100 percent (10/10), round-trip min/avg/max = 17/1364/1520 ms //注意并没有丢弃任何一个报文，同时请关注最后的平均通信时间为 1364 ms、最大通信时间为 1520 ms，这段时间非常长，这说明数据被缓存下载，在下个时间段被转发出去，这就是整形的特点：虽然慢，但尽量少丢弃报文
  C-R1#show access-lists QOS
  Extended IP access list QOS
    10 permit ip host 10.1.10.1 any
    20 permit ip 10.1.10.32 0.0.0.15 any (10 matches) //被 ACL 命中的流量
  C-R1#show policy-map interface
   Ethernet0/0

   Service-policy output: P1

    Class-map: C1 (match-all)
     10 packets, 8140 bytes
     5 minute offered rate 2000 bps, drop rate 0000 bps
     Match: access-group name QOS

```
  Queueing
    queue limit 64 packets
    (queue depth/total drops/no-buffer drops) 0/0/0    //没有报文被丢弃
    (pkts output/bytes output) 10/8140    //发送了 10 个报文
    shape (average) cir 8000, bc 32, be 32    //设置的转发速率，其他值系统自动匹配
    target shape rate 8000

  Class-map: class-default (match-any)
    177 packets, 23772 bytes
    5 minute offered rate 0000 bps, drop rate 0000 bps
    Match: any

    queue limit 64 packets
    (queue depth/total drops/no-buffer drops) 0/0/0
    (pkts output/bytes output) 177/23772
```

本案例实施完毕。

术语表

AAA：Authentication Authorization Accounting，认证、授权、审计。

ARP：Address Resolution Protocol，地址解析协议。ARP 是用来动态绑定高层 IP 地址到底层物理地址的一种 TCP/IP 协议。ARP 是通过二层广播方式寻找目的 IP 地址与目的 MAC 地址的对应关系的协议。

AS：Autonomous System，自治系统。

半双工：指在同一时间内只能在同一个方向进行的双向通信。一方在接收信息，而另一方在发送信息的通信，就是半双工。

报文：在数据通信领域，报文结构固定，头部定义了目的地址，文本就是实际的报文，也可以包括表示报文中止的信息。

报文丢弃：将来自未知 VLAN 域的报文或广播报文丢弃的功能。通常为了防止未知报文或广播报文占有当前链路的带宽资源，提高业务传输的可靠性，需要使用报文丢弃功能。

背板：一种电子电路板，包括线路和插座。其他电路板或电路卡上的电子设备可以插入到线路和插座中。在计算机系统中，背板与母板同义或隶属于母板。

背板容量：背板为线路板和交换网板之间提供的所有高速链路带宽的总和。

备用交换机：Standby Switch，是主交换机的备份交换机，堆叠/集群中只有一台备用交换机。当主交换机发生故障时，备用交换机接管主交换机的所有业务。

波特率：传输线路上信号每秒变化的次数。一般来讲，传输线路只有两种信号状态，波特率就是每秒传输的码元符号的个数。底层传输技术会占用一些带宽，因此业务数据不一定按照线路额定的波特率进行传输。

CAR：Committed Access Rate，承诺访问速率。

区分服务：简称 DiffServ。DiffServ 是一个多服务模型，可以满足不同的 QoS 需求。应用程序在发出报文前，不需要通知通信设备，而且网络不需要为每个流维护状态。它根据每个报文指定的 QoS 来提供特定的服务，包括进行报文的分类、流量整形、流量监管和排队。主要实现技术包括 CAR 和队列技术。

CIDR：Classless Inter Domain Routing，无差别域间路由。

CIR：Commit Information Rate，承诺信息速率。

CLI：Command Line Interface，命令行接口。

DAD：Duplicate Address Detection，重复地址检测。

代理 ARP：代理地址解析协议。通过该协议，中间设备（例如一台路由器）可以代表一个终端节点来向主机发送 ARP 响应报文。在速度较慢的 WAN 链接中，使用该协议可以减少带宽的使用。

带宽：模拟通信系统中最高频率和最低频率之间的差，或数字通信系统的数据传输能力。

例如表示信息流的速率,单位是 bps(bit/s)。

单播:也称为定点传送。信息被送往唯一的一个目的地。

单通:一种通话故障。在传统电话通信中,通话的双方中有一方不能听到对方的声音。在数据通信中,表示某方向报文发送和接收正常,而另一方向报文收发异常,导致两端通信异常。

单纤单向通信:为了保证报文分析服务器的数据安全,设备的光接口单板提供了单纤通信功能。用户通过使能发送端单纤通信命令,可以完成两个端口单根光纤单向通信。通过该功能可以实现分析服务器只接收报文,不外发报文,从而保证了分析服务器的数据安全。

DC:Data Center,数据中心。

点到点:两个站点共享一条传输路径的配置方式。

电气和电子工程师学会:采用会员制的组织,总部位于纽约,负责制定和发布技术标准、出版科技期刊。该标准组织主要制定二层协议。

地址掩码:表明地址大小的数值,同地址一起配置。"1"表示地址的"链路"部分,"0"表示"节点"部分。位掩码用于从 IP 地址中选取位数,用于子网寻址。掩码长度为 32 位,分别匹配 IP 地址的网络部分和一个或者一位或者多位本地(主机)部分。

EGP:Exterior Gateway Protocol,外部网关协议。

防火墙:防火墙类似于建筑大厦中用于防止火灾蔓延的隔断墙,是一个或一组实施访问控制策略的系统。它监控可信任网络(相当于内部网络)和不可信任网络(相当于外部网络)之间的访问通道,防止外部网络的危险蔓延到内部网络中。防火墙作为一种网络安全设备,包括硬件和软件。根据指定的策略过滤或拒绝流经的数据,可工作于 OSI 七层模型的任何一层。

访问控制列表:Access Control List,ACL。ACL 列举出每个用户访问某个目标时所拥有的权限的表格,以便于操作系统进行接入控制。在通信设备上,ACL 是一种通过报文的属性(如报文 IP 地址等)来识别特殊种类报文的流量过滤器。

封装:分层协议所采用的一种技术,底层协议收到来自上层协议的消息时,将该消息附加到底层帧的数据部分。在交付 A 协议的报文时,报文具有完整的 A 报头信息,作为数据由 B 协议进行承载。封装 A 协议的报文依次带有 B 协议报头、A 报头、A 协议作为本身数据进行承载的信息。值得注意的是,A 可以和 B 相同,例如,IP 协议承载 IP 协议报文。

服务质量:通信系统或信道的常用性能指标之一。在不同的系统及业务中其定义不尽相同,包括抖动、时延、丢包率、误码率、信噪比等。用来衡量一个传输系统的传输质量和服务有效性,评估服务商满足客户需求的能力。

GRE:通用路由封装。GRE 提供了将一种协议的报文封装在另一种协议报文中的机制,是一种隧道封装技术,使报文可以通过 GRE 隧道透明的传输。

根端口:在非根交换机上,离根交换机最近的端口叫作本设备的根端口。根交换机没有根端口。

广播:将数据报发送到网络中所有主机的通信方式。

广播地址:在计算机网络中,广播地址是一个允许信息发送给网络中所有节点的网络地址,而不是只将信息发送至一个特定的网络主机。

广播风暴:当一个节点向网内其他节点以广播方式发送错误的消息时,多个节点同时向其发出作为响应的信息帧，造成网络上大量信息帧拥塞的现象。广播风暴会以指数速度发展，使路由器超负荷运转，整个网络在很短时间内就会陷于瘫痪。

广播域：广播域是指一组网络节点，这些节点接收组内任何设备发送的广播报文。广播域也指设备上负责转发组播帧、广播帧和未知目的帧的所有接口的集合。

核心层：核心层是网络的高速交换主干，提供高速转发通信。核心层提供高可靠、高吞吐率、低延时性的传输主干。核心层设备需要具备优良的冗余性、容错性、可管理性、适应性，支持双机冗余热备份或负载均衡技术。在实际组网中，核心层包括由 NPE 和骨干路由器组成的 IP/MPLS 骨干网络。

汇聚层：汇聚层是接入层和核心层的"中介"，负责为接入层提供聚合与转发功能。汇聚层一方面处理来自接入层设备的所有通信量，同时提供到核心层的上行链路。与接入层设备相比，汇聚层设备需要更高的性能、更少的接口和更高的交换速率。

IGP：Interior Gateway Protocol，内部网关协议。

IS-IS：网络设备（路由器）使用的一种中间系统到中间系统协议。这种协议在以数据包为基础的网络上确立传送数据报文或数据包的最佳路径，即所谓的路由选择。

交换机：交换是按照通信两端传输信息的需要，用人工或设备自动完成的方法，把要传输的信息送到符合要求的相应路由上的技术统称。广义的交换机就是一种在通信系统中完成信息交换功能的设备。

尽力而为：网络技术特性之一，不提供链路层的可靠性。IP 协议在支持尽力而为发送机制的硬件上正常运行，这是因为 IP 协议不需要所运行的网络提供可靠性。UDP 协议向应用程序提供尽力而为发送服务。

距离矢量路由：采用分布式最短路径算法的一类路由更新协议。采用分布式最短路径算法的每台路由器向邻居发送清单，包括它所能到达的网络以及到达每个网络的距离。

快速生成树协议：该协议规范在 IEEE 802.1w 中有详细的描述。

LACP 抢占：在 LACP 静态模式下，如果链路聚合组中的某条活动链路出现故障，系统会从备用链路中选择优先级最高的链路替代故障链路，经过一段时间，被替代的故障链路恢复正常，而且该链路的优先级又高于替代自己的链路，在这种情况下，恢复正常的故障链路就可以回切到活动状态，而替换链路重新被切换为备份链路，这就是 LACP 抢占。

链路聚合组：将若干条物理链路捆绑在一起所形成的逻辑链路称为链路聚合组或者 eth-Trunk。

MAN：Metropolitan Area Network，城域网。

NAPT：Network Address and Port Translation，网络地址和端口转换。

NetStream 采样：针对样本报文进行流信息统计。

NQA：Network Quality Analysis，网络质量分析。

OSPF：开放式最短路径优先。开放式最短路径优先是在 Internet 团体中作为 RIP（路由信息协议）的后继者而被提出的链路状态分层 IGP 路由选择算法。OSPF 具有最低代价路由选择、多路径路由选择和负载平衡等特点。

PAP：Password Authentication Protocol，密钥认证协议。

PPP：Point-to-Point Protocol，点对点协议，主要用于在全双工的同/异步链路上进行点到点的数据传输。

PPPoE：PPP over Ethernet，以太网上的点对点协议，它提供了在以太网网络中多台主机连接到远端的宽带接入服务器上的一种标准。

QoS：Quality of Service，服务质量。

全双工：全双工也称双工，是指在同一时间内在两个方向同时进行的双向通信。

默认路由：指前缀匹配任何地址的路由，包括最长前缀路由中的零长度前缀。

认证：在多用户或网络操作系统中，系统对用户登录信息进行合法性检查的过程和方法。认证授权计费。

RFC：Request For Comments，征求意见稿，重要的互联网标准。

冗余：在系统或设备完成任务中起关键作用的地方，增加一套以上完成相同功能的功能通道、工作元件或部件，以保证当该部分出现故障时，系统或设备仍能正常工作，提高系统可靠性。

RSTP：Rapid Spanning Tree Protocol，快速生成树协议。

生成树协议：在局域网中用于消除环路的协议。

TFTP：简易文件传送协议。文件传送协议（FTP）的另一种小型简单协议形式。TFTP用于客户端和服务器之间不需要复杂交互的应用，它把业务限制在简单的文件传输上，不需要进行验证。TFTP可以存储在ROM中。

TTL：Time To Live，生存时间。尽力而为传输机制采用的一种技术，用于避免报文无限环回。发送方将TTL值设置为报文在网络中允许生存的最长时间。网络中的每台路由器在收到报文时，将TTL值减一；如果TTL值为零，将丢弃报文。

双归属：一种网络拓扑结构。设备通过两条独立的链路连接到网络上，两条链路互为主备链路，其中一条为主链路，另一条为备链路。应用在网络传输对安全性与可靠性要求较高的组网环境下。

SLA：Service Level Agreement，服务等级协定。

SSH：Secure Shell，安全外壳。

UDP：User Datagram Protocol，用户数据报协议。

VLAN：Virtual Local Area Network，虚拟局域网。

VPDN：Virtual Private Dial-up Network，虚拟专用拨号网。

VRRP：Virtual Router Redundancy Protocol，虚拟路由冗余协议。

WAN：Wide Area Network，广域网。

校验位：附加到二进制数列上对二进制数求和的检验位，最终结果恒为奇数或偶数。

循环冗余校验：将比特串看作带有二进制系数的多项式而计算出的一种帧校验序列（FCS）。循环冗余校验值为CRC多项式相除的余数。分组交换网络硬件负责计算CRC值并在传输开始前将CRC值附加到报文中。

虚拟交换机：通过软件模拟的一个在服务器内部用于虚拟机之间的交换机。由于是由软件处理的，随着虚拟机数量的增加，vSwitch会占用CPU资源而导致虚拟机性能下降，存在虚拟机流量监管、虚拟机网络策略实施以及vSwitch管理可扩展性等问题。

虚拟专用网：被定义为通过一个公用网络（通常是因特网）建立一个临时的、安全的连接，是一条穿过混乱的公用网络的安全、稳定的隧道。虚拟专用网是对企业内部网的扩展。虚拟专用网可以帮助远程用户、公司分支机构、商业伙伴及供应商同公司的内部网建立可信的安全连接，并保证数据的安全传输。

远程登录：Telnet，是 TCP/IP 协议堆栈中的标准终端仿真协议。远程登录应用于远程终端连接，通过它用户能像使用本地系统一样登录远程系统、使用资源。

载波侦听多路访问／冲突避免：CSMA/CA 采用主动避免碰撞而非被动侦测的方式来解决冲突问题，当设备欲发送数据帧且侦听到信道空闲时，维持一段时间后，再等待一段随机的时间依然空闲时，才提交数据。

帧：由数据、一个或多个地址以及其他协议控制信息组成的比特组，通常指链路层（OSI 的二层）协议数据单元。

直连邻居：两个邻居设备直接，没有任何中间设备，通过链路直接相连接，这两个设备彼此称为直连邻居。